# Dams Safety and Society

This book addresses current international practices applied for dam safety assessments by looking at a portfolio of dam safety projects in various developing countries (Armenia, Georgia, Tajikistan, Mauritius, Madagascar, Sri Lanka, Myanmar and Vietnam) spread across three continents (Europe, Africa and Asia). Safety assessment involved the review of 134 existing dams and comparison with the best international practices in developed countries. A large part of dam safety assessment involves understanding of dam hazards, standards applied in the design and maintenance, as well as expectation and social circumstances under which the dams have been designed and constructed in a particular country. For example, standards for design floods, ground investigation, selection of design soil parameters and design earthquakes etc. used are often either non-existent or inadequate, which could lead to an unsafe design. If there are no standards to be applied in dam design and construction, consultants are often under pressure from clients to come up with minimalistic investigation and designs which, a few years after dam construction, show signs of deficiencies. Very often countries have no regulations and standards for requirements that should cover the maintenance and operation of dams.

The book also describes the Portfolio Risk Assessment of Dams, which can be used as a tool by clients and the funding agencies to identify priority assessment and rehabilitation projects that consider societal and economic losses. It also demonstrates how the implementation of Emergency Preparedness Planning could significantly reduce the number of people at risk.

The book aims to help clients, consultants and funding agencies which are engaged in dam safety assessment projects in developing countries to focus on issues that are based on past lessons learnt.

# Dams Safety and Society

Ljiljana Spasic-Gril

## CRC Press
Taylor & Francis Group
Boca Raton  London  New York  Leiden

CRC Press is an imprint of the
Taylor & Francis Group, an **informa** business

A BALKEMA BOOK

Cover image: Ljiljana Spasic-Gril

First published 2023
by CRC Press/Balkema
Schipholweg 107C, 2316 XC Leiden, The Netherlands
e-mail: enquiries@taylorandfrancis.com
www.routledge.com – www.taylorandfrancis.com

*CRC Press/Balkema is an imprint of the Taylor & Francis Group, an informa business*

*Library of Congress Cataloging-in-Publication Data*
A catalog record has been requested for this book

ISBN: 978-0-367-33534-2 (hbk)
ISBN: 978-1-032-30510-3 (pbk)
ISBN: 978-0-429-32045-3 (ebk)

DOI: 10.1201/9780429320453

Typeset in Times New Roman
by codeMantra

For our dearest Mila

# Contents

# Preface

This book provides the most important dam safety issues and lessons learnt by working on dam safety assessments of 134 existing dams in eight countries, namely Armenia, Georgia, Tajikistan, Mauritius, Madagascar, Sri Lanka, Myanmar, and Vietnam, spread across three different continents (Europe, Africa and Asia).

The dams described in the book are for irrigation, potable water supply or hydropower. Some are over 2,000 years old (in Sri Lanka), but some others are only several years old (in Myanmar). The information presented covers a status being in place in these countries from the beginning till the end of the second decade of the 21st century.

The examples provided cover individual dams, as well as portfolio of dams. The book also discusses a method, the Portfolio Risk Assessment of Dams, that can be used as a tool by the clients (in Armenia, Georgia, and Sri Lanka) or the funding agencies (in Vietnam) to identify priority assessment and rehabilitation projects that can take into account societal and economic losses. It also demonstrates how the implementation of Emergency Preparedness Planning could significantly reduce the number of people at risk from flooding due to potential dam break.

By working on these projects, it was fascinating to witness how societies heavily rely upon dams; with population growth some dams became very close to populated areas, potentially presenting a risk from flooding or dam bursts.

A large part of dam safety assessments involves an understanding of dam hazards, standards applied in the design, construction, operation, and maintenance as well as expectations and social circumstances under which the dams have been designed and constructed in a country.

For example, standards for ground investigation, selection of geotechnical parameters for analyses, selection of design and safety check floods, selection of design earthquakes etc used in less developed countries are often either non-existent or unconservative, which often could lead to an unsafe design. If there are no standards to be applied in dam design and construction, consultants are often under pressure from the clients to come up with a minimal investigation and design costs, which after a few years of dam construction show signs of deficiencies. Very often, countries have no regulations and standards for requirements that cover maintenance, operation, or emergencies.

By working on these projects, I had the great opportunity to listen to the local colleagues and try to understand issues related to dams' safety, but also related to society, which made me realise that understanding the social background of the technical problems sometimes is one of the most important steps in understanding the problems themselves; they could largely impact design, maintenance, and operational decisions. Hence, this book mentions some of these socially driven dam safety aspects.

While on school or university holidays, my son Petar accompanied me on many of my overseas assignments; during our travels to faraway places, we had many opportunities to discuss local social issues and how dams affected the lives of local people. On many of these occasions, we discussed the possibility of writing a book together, which should have engineering dam safety elements, but would also cover some social perspectives.

In my career I worked on over 400 dams, in 56 countries, however, the examples given in the book are regarded to be typical where safety aspects of the dams have largely been driven by various social elements. These eight countries have been selected to demonstrate how some social and political circumstances led to similar decisions that affected dams' safety and dams' conditions.

It is hoped that, by highlighting the most critical aspects of dam safety and their links with the social requirements, this book will be useful not only to the practitioners and Dam Industry Policy makers working in the countries discussed in the book but also in other counties where there is no clear dam safety guidance for safety assessment of dams.

*A word from Petar Gril (the author's son)*

As a teenager and in my early twenties, I travelled to countries where my mother had dam projects and I saw, at first-hand, what life was like for people living in the nearby communities to dams. I met a lot of local people and had a chance to talk to them about different life topics, including how important water, energy and dams are to their communities. Those social aspects were the ones I was particularly interested in. I visited Sri Lanka, Tajikistan and Mauritius, the countries mentioned in this book.

When a few years ago my mother said that she was going to write a book about safety issues on some existing dams she worked on, it was only natural for me to ask her if I could contribute in any way. For this book, I gathered general information on a particular county, as well as information on total numbers of dams etc.

## SRI LANKA

My first experience of dams and the impact they have on societies was in Sri Lanka in 2003. That year I travelled to Asia for the first time as a 13-year-old boy and it enlightened me into understanding how water is integral for every society in the world and how dams have a vital role.

From the bustling streets of Colombo, to the long sandy beaches of Galle and temples in Kandy it became apparent to me of what this mystical country had to offer and the ways in which it encapsulated you with its smorgasbord of religions, cultures, and traditions, all in one special place.

After travelling around Colombo on tuk-tuk we embarked on our first journey outside the capital to an elephant sanctuary in Peradeniya, where I became acquainted with the Asian Elephant in all its glory. It was a beacon for Buddhist temples which were the backbone of society there like the ascent to Dambulla, where we were besieged by a tribe of macaque monkeys.

This was on our way to see Polonnaruwa Dam, which is one of the oldest in Asia dating back to over 2,000 years old. My mother had to inspect it and assess its safety and gauge whether it needed rehabilitation like the many other dams dotted about the island that pose risk to people.

This trip to Sri Lanka showed me how important water is to society now as it was a couple of 1000 years ago, and how dams were recognised as a reliable structure to provide that facility.

## TAJIKISTAN

In 2006, I accompanied my mother on a visit to the Central Asian Republic of Tajikistan; this journey would long live in my memory as this was one of the poorest places in the world that I had been to.

From a dams' perspective, the main difference with Tajikistan compared to Sri Lanka is that most of the dams in the republic are for hydropower as opposed to irrigation or water supply - the country depends upon the dams for electricity.

During the visit we were accompanied by local experts as they took us on some treacherous terrains and dirt roads at altitudes of over 3,000 m above sea level -we saw the tallest dams in the world – Nurek and Rogun (which was under construction at the time). After the breakup of the Soviet Union, Tajikistan was exposed to a decade of civil war, but all parties recognised the importance of the Nurek Dam, constructed in the 1970s; the dam was well protected and was not affected by the civil war.

For several days we also stayed with a local family, in a remote village at over 3,000 m elevation, where a small hydropower dam was under construction. The families in the village do not have running water nor electricity. It was interesting to see how people lived in these remote, mountainous areas of Tajikistan. The people there had truly little, but they still managed to carve out a sustainable life for themselves; I still remember and treasure their hospitality and our conversations in the evenings over dinners.

## MAURITIUS

In 2012, I had a chance to spend a month in the proverbial paradise - the island of Mauritius. This tropical haven is situated off the coast of Madagascar, near the Indian ocean islands of Seychelles and Reunion.

It was here that I realised how important one dam can be to the people living in areas nearby. There was a large community of people living at the foot of La Ferme Dam. The people used the reservoir to wash, play and for drinking water.

I have never seen a dam used for so many purposes before, I was now able to understand the impact that dams have on society around the world.

# Acknowledgements

I am thankful to the clients of the projects mentioned in this book for letting me use the information, which mainly came from my previously published papers on the projects.

During my work on many international dam projects, I had a chance to meet and work with many remarkable people, specialists in their fields – I learnt a great deal from them. I always regarded them as my close colleagues, which I have remained in touch with to date. Just to mention a few: Prof. Paul Marinos, Dr. Peter Mason, Dr. George Annandale, Alessandro Palmieri, Iftikhar Khalil and Satoru Ueda.

I am also grateful to many local experts for knowledge sharing, especially related to local organisations and procedures; our collaboration often went beyond technical discussions and involved social, cultural (Figures 0.1 and 0.2) or political topics, touched on during our meetings or meals together, which always helped to put into a context our technical recommendations and decisions. There are many colleagues from Armenia, Georgia, Tajikistan, Mauritius, Myanmar, and other countries that I could mention; I have stayed in touch with many of them. However, I feel that Mrs. Badra Kamaladasa, former Director-General of Irrigation Department in Sri Lanka, deserves to be singled out. We have worked together on the Dam Safety Project in Sri Lanka (Section 6), and on many occasions discussed technical and operational issues and how the dam safety training and legislation should be shaped in Sri Lanka. We since met at a few ICOLD meetings and remained in touch; Badra has written to me during the COVID-19 pandemic in 2020 to tell me exactly what I am trying to say here – how important our encounter, collaboration and friendship is to us.

A big thank you to my colleague Jason Manning from Arup in London, who helped me with the production of the maps.

I am so grateful to my friends who encouraged me to write the book, and to carry on with the writing; Ana Klasnja you have been the most persistent with the encouragement, which first started several years back, after me telling you my stories about the visit to Lake Sarez in Tajikistan (Section 5) and you arranging an exhibition in Ontario Science Centre in Toronto about my involvement in this project. Thank you.

I am also grateful to my son Petar Gril for being my companion on some of my business trips, for sharing his views about my work, encouragement to write and for his contributions to the book.

Finally, I would like to thank my husband Dragutin Gril for putting up with me being absent during my frequent business trips, many times to faraway places, sometimes with no telephone connections; I know that, while I was travelling, concerned about the safety of various dams, you were worried about my own health and safety during these trips; also, you have never seemed to show that you are bored from listening to my dam stories over and over again.

*Figure 0.1* Gisar fortress, Tajikistan - built around 550 BC by Cyrus the Great, the Persian emperor.

*Figure 0.2* Laykyun Sekkya Buddha, Myanmar, the third-tallest statue in the world at 116 m.

# Abbreviations

| | |
|---|---|
| ALARP | As low as reasonably possible |
| CEB | Ceylon electricity board |
| CFD | Computational fluid dynamics |
| DSAP | Dam safety action plan |
| DSHA | Deterministic seismic hazard assessment |
| DSP | Dam safety project |
| DS&RCP | Dam safety and reservoir conservation project |
| EPP | Emergency preparedness plan |
| ESIA | Environmental and social impact assessment |
| EWS | Early warning system |
| FSL | Full supply level |
| GBR | Geotechnical baseline report |
| GDP | Gross domestic income |
| GLOF | Glacial lake outburst flood |
| GNI | Gross national income |
| HPP | Hydropower plant |
| H&S | Health and safety |
| ICOLD | International commission on large dams |
| ID | Irrigation department |
| IDCDP | Irrigation and drainage community development project |
| IDNDR | International decade for natural disaster reduction |
| masl | metres above sea level |
| MASL | Mahaweli authority of Sri Lanka |
| MCM | Million cubic meters |
| MDE | Maximum design earthquake |
| MS | Monitoring system |
| OBE | Operating basis earthquake |
| O&M | Operation and maintenance |
| PFMA | Potential failure mode analysis |
| PGA | Peak ground acceleration |
| PMF | Probable maximum flood |
| PMP | Probable maximum precipitation |
| PRA | Portfolio risk assessment |
| PSHA | Probabilistic seismic hazard assessment |
| QRA | Quantitative risk assessment |
| RAP | Resettlement action plan |

| RCC | Roller compacted concrete |
| SEE | Safety evaluation earthquake |
| SGI | Supplementary geotechnical investigation |
| SMA | Strong motion accelerometers |
| SNIP | Soviet design norms |
| SSSHA | Site-specific seismic hazard |
| ToR | Terms of reference |
| WB | World bank |

# About the author

**Ljiljana Spasic-Gril** has 40 years experience in the feasibility, design, dam safety assessment, construction and commissioning of dams and reservoirs in Europe, Africa, the Middle East, Central, South and East Asia, Australasia, North and South America. She has worked on around 400 dam projects in over 56 countries. Ljiljana has extensive experience on dam safety projects worldwide and has been involved in dam safety projects in the UK (8 dams), Sri Lanka (30 dams), Russia (one dam), Armenia (65 dams), Georgia (6 dams), Ethiopia, Vietnam (450 dams) and Central Asia (over 30 dams). The dam safety project in Central Asia included safety assessment of the Nurek dam, a 300 m high embankment dam, currently the second tallest dam in the world, and the Rogun Dam in Tajikistan when constructed to be the world's tallest dam (335 m tall). Ljiljana has wide experience in managing the activities of multi-disciplinary teams comprising expats and local engineers; she has been involved in developing major schemes from inception through to implementation funded by the World Bank, Asian Development Bank, Islamic Development Bank, African Development Bank, EBRD, EIB, KfW, DfID and other international and UK funding agencies. Ljiljana has also been appointed as a Dam Expert to UNESCO World Heritage. She is author of 18 papers (6 presented at ICOLD conferences), a Chapter on Dams in the book "Engineering Geology and Geomorphology of Glaciated and Periglaciated Terrains" published by the Geological Society, Engineering Geology Special Publication 28. She is the UK representative to the Seismic Technical Committee of International Committee for Large Dams (ICOLD) and Permanent Visiting Professor at South Bank University, London (Dam Engineering).

# Chapter 1

# Introduction

## 1.1  BACKGROUND ON DAM SAFETY

Dams are among the greatest human achievements and their history dates back to 3000 BC. They impound reservoirs that are used for water supply for irrigation, consumption, industrial use, hydropower, navigation, or flood protection. About 30% of dams are now multipurpose.

There are over 800,000 dams in the world; more than 57,000 dams have been classified as large dams (as per ICOLD classification). China has the world's largest number of large dams (over 23,000 large dams), followed by the United States (over 9,200 large dams) and India (over 5,200 large dams).

Dams have played a key role in fostering rapid and sustained agricultural, rural and urban development. In the United Kingdom, for example, to meet the demands of industrial growth, the number of dams built surged during the 19th century, and there are approximately 570 of them in the country today (dated back to that period).

While providing great benefits, dams could also pose potential risks to the neighbouring population, properties and infrastructure, as well as the natural environment. Also, many dams constructed before or during the 1960s and 1970s have reached their design life, which, in the past, was assumed to be 50 years.

Although a dam failure has a very low probability of occurrence, there are severe impacts on lives and properties when these failures occur. Based on historic data, the average annual failure rate for dams is $10^{-4}$ and the probability of death per person living in a potential inundation area is in the range of $10^{-7}$ to $10^{-6}$.

According to statistics (ICOLD Bulletin 99[1]), 2,000 dams failed since the 12th century, 200 dams failed in the 20th century (outside China), and more than 17,000 people died because of dam failures.

Based on ICOLD Bulletin 90, 80% of total dam failures were within embankment dams, which also caused 70% of total dam fatalities (see Table 1.1).

A hazard is defined as *"a Potential of a situation for harm to persons or property"*

A risk is defined as *"a severity of hazard * its probability of occurrence"*.

As the world's stock of dams gets older, and as the population living in potential inundation zones increases, both the rate of dam failures and the probability of fatalities will increase unless effective dam safety measures are implemented, including proper dam inspection, safety assessment, operation and maintenance, implementation of safety rehabilitation works, and emergency preparedness.

DOI: 10.1201/9780429320453-1

*Figure 1.1* Teton dam, downstream view after failure in 1976 resulting in 11 fatalities.

*Table 1.1* Dam failure statistics

|  | Concrete dams | Embankment dam |
| --- | --- | --- |
| % of total dam population | 30 | 70 |
| % of total failures | 20 | 80 |
| % of total dam fatalities | 30 | 70 |
| Most frequent cause of failure | Foundation overstress | Internal or external erosion |
| Type of failure | Sudden | Gradual |

When discussing a reduction of the "Risk of a Dam Failure", two elements are important:

- To minimise the probability of failure;
- To minimise the consequences of failure

To minimise the probability of failure of an existing dam, the following is typically required to be undertaken:

- Keep design, construction and operation and maintenance records;
- Monitor dam performance with instrumentation;
- Undertake periodic inspection and safety assessment;
- Apply a high standard of maintenance;
- Use appropriate operation and maintenance of equipment;

- Identify deficiencies through inspections;
- Undertake remedial works as soon as deficiencies are noted;
- Undertake a dam break analysis to evaluate impact of a dam failure on the down-stream population and produce Emergency Preparedness Plan (EPP);
- Check/review corrective actions identified in the EPP

## 1.2   THIS BOOK

Because of the increased population and demand for water, it is critical to keep the existing dams safe so that they can be used over a prolonged period. With population growth and expansion, it is becoming extremely difficult to resettle people in order to build new dams. Therefore, the emphasis is on keeping the existing dams as safe as possible over a long period.

Due to the population growth and dams' ageing, dam safety has been given significant importance in many countries. Dams are structures that, if fail, can pose a significant risk to downstream population, environment, and economy. This risk could be much larger than the risk due to the failure of any other structure. Therefore, maintenance of the dams and planning for downstream protection have become mandatory. In addition, over the past several decades, there is also knowledge gained on the performance of dams in extreme floods and earthquakes, which calls for a re-assessment of dams' performance in extreme events and rehabilitation to ensure their safety.

With an increasing number of ageing dams, dam safety has been seen to be an important factor for safeguarding investments in infrastructure as well as human lives and properties of the people living downstream of the dams. There is a demand that dam owners ensure their dams are operated and maintained in a safe and proper manner in accordance with accepted safe practice and relevant legislation.

For this book, the author selected 134 existing dams in eight countries on different continents with different social and political backgrounds. The selected dams were part of various dam safety projects in which the author was involved.

Although some dams referred to in this book are very old (Polonaruwa dam in Sri Lanka, about 2,500 years old, and La Ferme dam in Mauritius, 120 years old), most dams discussed in the book were built less than 50 years ago, some less than 10 years ago, but with significant problems that can cause safety risks.

Most of the dams selected are large dams ($H > 15$ m) according to the ICOLD classification.

Although ICOLD propagates certain dam safety standards and provides guidelines, differences in the design standards applied or dams' operation and maintenance procedures are still wide.

In analysing the safety of 134 existing dams in eight countries, the same international dam safety standards have been applied, largely grouped around the following aspects:

- dam safety inspection
- robust technical assessment of dam safety including safety to floods, landslides, earthquakes, construction details, etc;

- risk to downstream population due to dam breach and dam hazard classification;
- risk mitigation measures: technical and non-technical that include monitoring, safe dam operation and emergency planning and response

When discussing the above dam safety aspects, some social and political aspects which influenced dams' safety have also been mentioned.

## NOTE

1  ICOLD Bulletin 99 - Statistical analysis of dam failures.

# Chapter 2

# Main aspects in dam safety assessment and principles and concepts applied

## 2.1 DATA COLLECTION AND ASSIMILATION

It is important to undertake data collection before a dam inspection, as this will indicate areas where more attention during inspection might be needed. The data should comprise design drawings, construction - as-built records, records of inspections, maintenance, surveys, instrumentation monitoring, and rehabilitation works undertaken.

When dealing with existing dams, especially the ones in developing countries, very often only limited information exists regarding the construction of dams, and typically, the information available is the design data rather than construction records. That was typically the case with dams in the countries described in this book.

Often, for many of the smaller dams, no records were found at all. Sometimes it is possible to undertake an archive search.

One other important point is that, for example, in Tajikistan (see Section 5), Nurek and Rogun dams were originally designed by Soviet Consultants, when Tajikistan was a part of the Soviet Union. After the break-up of the Soviet Union, a lot of archive data on the dams remained with the previous consultants, which complicated maintenance procedures.

## 2.2 APPROACH TO DAM SAFETY EVALUATION

Before undertaking a dam safety evaluation, it is useful to understand the following:

- dam design life
- Safety criteria to be used for assessment of dam design life
- Dam Risk Classification
- Potential Failure Modes

These are separately addressed in the sections below.

### 2.2.1 Information on dam design life

The service life of a well-designed, well-constructed and well-maintained and monitored embankment and concrete dams are typically expected to be 100 years. However, many older dams have been designed to a design life of 50 years.

DOI: 10.1201/9780429320453-2

The life span of hydromechanical steel structures, electromechanical equipment and control units is typically shorter than that of the main civil/structural components and are specified by the suppliers, who also provide instruction manuals describing operation and maintenance.

It has to be recognised that there is a direct relationship between dam safety and its life span, i.e. if the dam is unsafe its life span has expired.

It can also be noted that the useful life of a dam can be affected by an increase in the sediment transport of the river, which could cause an increase in the sediments retained in the reservoir, thus reducing the live storage of the reservoir, if sediment mitigation measures are not appropriate or not implemented.

### 2.2.2   Safety criteria for assessment of the design life of dams

The life span of any dam is as long as it is technically safe and operable. In view of the high damage potential of large storage dams, the safety has to be assessed based on an integral safety concept, which includes the following elements:

1. Structural safety (main elements: geologic, hydraulic, and seismic design criteria;
2. Safety monitoring (main elements: dam instrumentation, periodic inspections and safety assessments by dam experts, etc.).
3. Operational safety (main elements: reliable rule curves for reservoir operation under normal and extraordinary (hydrological) conditions, training of personnel, dam maintenance, sediment flushing, engineering backup, etc.).
4. Emergency planning (main elements: emergency action plans, water alarm systems, evacuation plans, engineering backup, etc.).

Therefore, as long as the proper handling of these safety issues can be guaranteed according to this integral safety concept, a dam can be considered safe. With the number of people living in the downstream area of a dam and the economic development, the risk pattern may change with time, calling for higher safety standards to be applied to the project.

### 2.2.3   Dam risk classification

Dam Risk Classification takes into account the following:

- dam height ($H$) and volume ($V$)
- downstream population at risk and
- potential economic losses.

Dam Risk Classification can be undertaken using ICOLD guides. ICOLD bulletin 157[1] defines a large dam as dam with:

- $H > 15$ m, or
- $10$ m $< H < 15$ m, Storage $> 1$ million m$^3$
- $10$ m $< H < 15$ m, Crest length $> 500$ m

- $10 \text{ m} < H < 15 \text{ m}$, Spillway capacity $> 2,000 \text{ m}^3/\text{s}$
- $10 \text{ m} < H < 15 \text{ m}$ with unusual characteristics in dam type or foundations

All other dams are classified as small dams.

ICOLD Bulletin 72[2] provides the following Risk Classification (Tables 2.1 and 2.2):

Table 2.1  Risk factors

| Risk factor | Extreme | High | Medium | Low |
|---|---|---|---|---|
| | Contribution to risk (weighting points) | | | |
| Capacity (106m3) | >120 (6) | 120–1 (4) | 1–0.1 (2) | <0.1 (0) |
| Height (m) | >45 (6) | 45–30 (4) | 30–15 (2) | <15 (0) |
| Evacuation requirements (no of persons) | >1,000 (12) | 1,000–100 (8) | 100–1 (4) | None (0) |
| Potential downstream Damage | High (12) | Moderate (8) | Low (4) | None (0) |

Table 2.2  Risk Classes

| Total risk factor classification factor | Risk class |
|---|---|
| 0–6 | I |
| 7–18 | II |
| 19–30 | III |
| 31–36 | IV |

It should be noted that, in case of existing dams, other factors such as the availability or lack of construction and maintenance records, processed instrumentation and surveillance records, previous safety evaluations, new planned downstream development, etc. may affect the risk factors.

## 2.2.4   Potential failure modes

a.   General background

Potential Failure Modes Analysis (PFMA) is a method of analysis where particular faults or initialling conditions are postulated, and the analysis reveals the full range of effects of the faults or the initiating condition of the system.

This PFMA is based on the processes developed by the United States Bureau of Reclamation (USBR) and the United States Federal Energy Regulatory Commission (FERC) for dam safety assessment and management. The process is described in the FERC Engineering Guidelines[3] and is used for risk-informed decision-making.

Identifying, fully describing, and evaluating site-specific potential failure modes (PFM) are the most important steps in conducting a risk analysis. The PFM can only be identified after thoroughly reading all relevant background information on geology, dam and ancillary structures design, flood and earthquake loading, operation, performance, and monitoring documentation.

The ideal process for identification must go loading by loading (flood, earthquake, etc.) and feature by feature (foundation, downstream slope, embedded pipes, spillway, control of gates, etc.). The failure description must encompass the full sequence of events from initiation (cause) through to the realisation of ultimate failure and uncontrolled release of the reservoir and/or significant loss of operation/control.

This analysis helps identifying structural related PFM (e.g. piping) not covered by the commonly used analysis methods (e.g. slope stability)

After identifying the PFM, a list of factors that make each failure mode more and less likely to occur must be detailed, including results from monitoring reports, design details. This allows a better understanding of the process of each mode.

Each PFM must be discussed to identify risk reduction actions (surveillance/monitoring, investigations, remediation activities)

While identifying the PFM, the following questions must be addressed.
• How could this dam fail?
• Are the identified potential modes of failure recognised and appropriately monitored (visual surveillance or instrumental monitoring)
• What actions can be taken to reduce dam failure likelihood or to mitigate failure consequences?

b.  Categorisation of the PFM

Once identified and described the PFM must be classified as credible and significant by answering the following questions
• Is the PFM credible?
• Is the PFM significant?

The category usually considered are as follows.
• Category I – Highlighted Potential Failure Mode. PFM of greatest significance considering need for awareness, potential for occurrence, magnitude of consequence and likelihood of adverse response are highlighted
• Category II – PFM considered but not Highlighted. These are judged of lesser significance and likelihood.
• Category III – More information or analyses are needed in order to classify these PFM. Action may be highlighted
• Category IV – PFM ruled out because the physical possibility does not exist, or the PFM is clearly so remote a possibility as to be non-credible.

c.  Likelihood of failure – Event tree analysis

Category I PFMs must be studied to estimate the annual probability of failure following this PFM. This can be derived from event tree analysis and is based on the review of the existing background information and completed with general knowledge from similar failure cases studied for unknown conditions parameters.

d.  Consequences evaluation

The other aspect of risk analysis is the consequences evaluation.

The consequences in case of the uncontrolled release of the reservoir water must be evaluated. Different dam break scenarios must be computed in order to prepare inundation maps that help evaluate the consequences.

Consequences will be evaluated in terms of population at risk and economical losses (nature of building located in the inundation area).

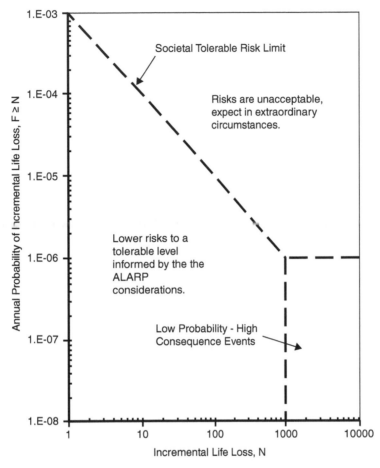

*Figure 2.1*  **F-N chart for societal incremental risk.**[42]

e.  Risk analysis

The Outputs of the Two Last Steps – annual probability of failure and consequences in case of failure are plotted on an f-N chart shown in Figure 2.1. The Chart displays the estimated probability distribution loss for a reservoir encompassing all failure modes and all population exposure scenarios for the incremental flood risk.

## 2.3  SITE INSPECTION

The field inspections are generally carried out by dam specialists accompanied by local technical staff and, where possible, by the operators.

The USBR guide on Safety Evaluation of Existing Dams[4] provides a good checklist for the onsite examination of embankment and concrete dams, appurtenant structures, access roads, and the reservoir rim.

The Inspection typically covers:

- Dam Access – Access to the dam site is important for ongoing surveillance and maintenance of the dam. Access roads need to be accessible in a reservoir emergency (see Figure 2.2 below) and so need a good standard of protection from flooding and from other hazards such as rockfalls and landslides. The roads need to be designed with sufficient width and turning space for the largest vehicles envisaged as being needed during inspection, operation, maintenance, or emergency response.

  The site access is likely to be needed to both abutments of the dam as well as the toe of the dam on both sides of the watercourse. Vehicle bridges are therefore typically required over obstructions and structures such as the spillway. As flood storage reservoirs only impound intermittently access should be provided to both the up and downstream toe of the dam.
- Reservoir Rim – the rim needs to be inspected; this is especially important in a complex geological environment, where landslides have occurred in the past or are likely to be triggered by the reservoir operation
- The Dam – this includes:
  - main dam
    - Upstream Face – slope protection, vegetation, settlements, debris, burrows, etc,
    - Downstream Face – signs of movement, seepage, and wet areas, vegetation, trees (see Figure 2.3), condition of slope protection, burrows, etc
  - Abutments – seepage, cracks, vegetation, signs of movement
  - Crest – surface cracks, settlement, camber

*Figure 2.2* **Access road to a dam in Madagascar – it cannot be used after heavy rains.**

*Figure 2.3* Inspection of La Ferme dam in Mauritius – mature trees are right at the downstream toe of the dam.

- Draw-Off Works – these need to be inspected; if concerns are at the upstream end of the draw-off works, inspection by divers is recommended
- Spillway (main and emergency)
  - Approach Canal – observe for plants or clogging with logs, trash, or construction debris; side banks for traces of slips; bottom of the canal;
  - Weir – check the overall condition of the concrete surface; check for cracks or other failures; check the crest and sides;
  - Weir Channel – check the side walls and the bottom for cracks and debris or vegetation;
  - Bridge Over the Spillway – if there is a bridge over spillway check its condition
  - Chute – check for debris, vegetation, or collapsed earth;
  - Stilling Basin – check for cracks, siltation;
- Monitoring & Instrumentation
- Hydromechanical and electrical equipment

If a dam is a part of a hydropower plant (HPP), also to be inspected are:

- HPP Pressure Tunnel
- HPP Penstock-
- HPP Underground Tailrace Gallery – The gallery could not be inspected as the HPP was in operation.
- HPP Tailrace Channel – No inspection of the bottom and walls was done as the HPP was in operation.
- HPP Powerhouse – the building, the power generating and mechanical equipment located in the powerhouse are in good condition.

## 2.4  SAFETY EVALUATION OF DESIGN, CONSTRUCTION, AND OPERATION

The safety evaluation includes studies and reviews which must determine if a dam is safe. These studies and reviews typically include the following:

- Hydrology, flood routing, design, and safety check floods
- Dam breach assessment
- Sedimentation,
- Review of geological, geotechnical, and geophysical conditions
- Seismicity
- Reservoir Basin
- Embankment Dams – Stability, settlements, and seepage analysis,
- Concrete Dams
- Spillway
- Outlet works
- Instrumentation, Operation, and Maintenance

In the sections below, a more detailed description of the work involved is presented. Provided also are typical international standards that are used for dam safety projects funded by International Funding Agencies.

### 2.4.1  Hydrology, flood routing, design, and safety check floods

a.  Catchment Hydrological and Hydraulic Modelling

The hydrological methods described in ICOLD Bulletin 170[5] are typically used to determine effective rainfall, peak flows from the catchments, and hydrographs. A variety of methods can be applied and the sensitivity to different estimates is considered with cognisance of the strengths and limitations of each method. A range of events of different return periods will be required including but not necessarily limited to the reservoir design flood, safety check flood, and flood events required for the assessment of downstream flood risk, design of downstream river flood defences, or flood risk of existing banks and defences. It is also anticipated that a range of storm events will need to be considered during different stages of construction to enable safe planning of the construction.

A selection of inflow hydrographs with different storm duration is typically generated for the design and safety check flood events. These hydrographs are routed through a 1D numerical reservoir hydraulic routing model (such as HEC-RAS) to determine the critical storm duration that results in the highest reservoir Stillwater level and peak outflow with these values used for design. For the assessment of downstream river flood risk and design of downstream flood defences a range of storm durations should be assessed.

b.   Glacial Lake Outburst Floods

Glacial Lake Outburst Floods (GLOFs) have been defined by the *United Nations* Platform for Space-based Information for Disaster Management and Emergency Response (*UN-SPIDER*) as "the sudden release of a significant amount of water retained in a glacial lake". This hazard is specific to any proposed dam location that is within a glacial catchment containing a glacial lake.

c.   Impacts of Climate Change

Climate change is already leading to changes in the intensity and frequency of events that can impact dam infrastructure; these impacts are anticipated to continue increasing and intensifying in future climate conditions. An understanding of potential climate change impacts is key to ensure that dam infrastructure is resilient to current and future climate impacts. This section provides a high-level overview of existing guidance. A detailed analysis would be required to assess impacts at specific locations.

**Existing guidance**

At the international level, there are a few limited examples of guidance and studies assessing climate change impact on dams. For example, in the UK there is specific guidance on the use of climate data to assess impacts on dams and reservoirs,[6] and there are examples of academic literature addressing and suggesting approaches to estimate potential climate change impacts on dams.[7] In addition, general guidance for designing climate-resilient infrastructure is applicable to dams.[8]

**Approach to assessing climate change impacts**

When assessing climate change impacts on infrastructure there are a set of key aspects to be considered in the analysis of climate change data. These are summarised below:

Climate Variable and Characteristics – An understanding of key climate variables for the impact analysis (e.g. precipitation and temperature) and the relevant characteristics (e.g. occurrence of extreme events or seasonal patterns).

• Time Period – The time period usually covered by climate models is 1950–2100. Climate change data should be analysed for the future period of interest in line with the design life of the infrastructure and key replacement or maintenance cycles.

• Scenarios – It is important to consider multiple (at least two) scenarios

• Climate Model Data and Analysis – It is important to consider which is the most suitable data source for the specific impact analysis and whether any additional processing or analysis is needed (e.g. statistical downscaling to obtain insight on change in extreme weather events at the local scale)

• Uncertainty – climate change projections are inherently uncertain, uncertainty in climate model outputs and scenarios should be assessed and represented.

• It is recommended to develop extra scenarios to take into consideration the impact of climate change by increasing these hydrographs by an agreed percentage (%).

d.   Design and safety check floods

For each dam, an appropriate design and safety check flood should be determined based on the hazard of the dam. ICOLD Bulletin 82 defines the design floods and safety check floods as follows:

- Safety Check Flood – It is considered acceptable practice for the crest structure, waterway, and energy dissipator to be on the verge of failure but to exhibit marginally safe performance characteristics for this flood condition.
- Design Flood – Strictly representing the inflow that must be discharged under normal conditions with a safety margin provided by the freeboard.
- Selection of such floods varies between different national standards. These floods should be selected based on the dam hazard classification or consequences that would result from a dam failure. Two potential methods may be taken:

*A prescriptive standards-based approach* where a dam hazard category is assigned based on criteria such as dam height, stored volume, population at risk downstream, and economic consequences. The design standards then dictate the design and safety check floods depending on the hazard. This approach is described in ICOLD Bulletins 82[9] and 125[10] and 157 (Endnote 1) for small dams. Some example flood standards are provided in Figure 2.4.

---

**Guidelines of the US Army Corps of Engineers**

**Table 3 - Size classification**

| Category | Reservoir capacity (hm³) | Height of the dam (m) |
|---|---|---|
| Small | from 0.62 to 1.23 | from 7.6 to 12.2 |
| Intermediate | from 1.23 to 61.5 | from 12.2 to 30.5 |
| Large | ≥ 61.5 | ≥ 30.5 |

**Table 4 - Hazard potential classification**

| Category | Loss of life (Extent of development) | Economic loss (Extent of development) |
|---|---|---|
| Low | None expected (No permanent structures for human habitation) | Minimal (Undeveloped to occasional structures or agriculture) |
| Significant | Few (No urban developments and no more than a small number of inhabitable structures) | Appreciable (Notable agriculture, industry or structures) |
| High | More than few | Excessive (Extensive community, industry or agriculture) |

**Table 5 - Recommended safety standards**

| Hazard | Size | Safety standard |
|---|---|---|
| Low | Small Intermediate Large | 50-yr to 100-yr flood 100-yr flood to 50 % of the PMF 50 % to 100 % of the PMF |
| Significant | Small Intermediate Large | 100-yr to 50 % of the PMF 50 % to 100 % of the PMF PMF |
| High | Small Intermediate Large | 50 % to 100 % of the PMF PMF PMF |

*Table A29-1 USA – IDF requirements for dams using a prescriptive approach*

| Hazard potential classification | Definition of hazard potential classification | Inflow design flood |
|---|---|---|
| High | Probable loss of life due to dam failure or misoperation (economic loss, environmental damage, or disruption of lifeline facilities may also be probable, but are not necessary for this classification) | PMF1 |
| Significant | No probable loss of human life but can cause economic loss, environmental damage, or disruption of lifeline facilities due to dam failure or misoperation | 0.1% annual chance exceedance flood (1,000-year Flood)2 |
| Low | No probable loss of human life and low economic and/or environmental losses due to dam failure or misoperation | 1% annual chance exceedance flood (100-year Flood) or a smaller flood justified by rationale |

177

(Note: PMF- Probable Maximum Flood)

*Figure 2.4* **Example flood standards (ICOLD Bulletins 82 and 125).**

(Note: PMF- Probable Maximum Flood)

Alternatively, *a risk-based approach* may be taken where the incremental damages caused by a failure in a range of events are considered along with a comparison of flood damages not resulting from a dam failure. This aims to achieve a risk that is As Low as Reasonably Practicable (ALARP). Assessing such risk requires consideration of the type of dam proposed as, for example, embankment dams are more vulnerable to overtopping than concrete.

The chosen method, design, and safety check floods are to be agreed upon with the client. Where there is more than one reservoir within a catchment (a reservoir cascade) the risk of a failure of one reservoir resulting in an unsafe condition in another should be considered.

A climate impact study should be undertaken and appropriate sensitivity testing shall be carried out to determine the consequences for the performance of the reservoirs. This should consider three key areas:

- The future adequacy of spillway and outlet works
- Impact on downstream flood risk
- Other impacts where dam is intended for multiple purposes

The whole life cost of the dams should be assessed considering the need for future climate change adaptations, either providing additional capacity where appropriate or allowing for ease of future modification with the decisions based on the lowest whole life cost.

## 2.4.2 Dam breach assessment

A dam breach assessment is typically required for a dam, for the following purposes:

- To provide dam breach inundation mapping for development planning
- To inform determine the consequences of a dam breach and therefore the selection of the reservoir design and safety check floods
- For emergency planning and development of the EPP

The dam breach assessment should consider the potential for different types and locations of failure for example due to piping or overtopping, at structures or the highest part of the dam with different breach hydrographs being developed. Different methods should also be considered with the sensitivity to different events considered. Different flood events will need to be considered with, as a minimum a sunny day and extreme flood scenarios, additional flood events may need consideration when assessing incremental flood damages. In each case, the flood extents should be plotted and the time of travel for the flood wave noted for settlements and other key locations. The approach to developing dam breach assessments should be is described in ICOLD Bulletin 111.[11]

The dam breach assessment will be used to inform an EPP, the draft plan should be developed for consultation with all relevant stakeholders including government ministries, local communities, and emergency services. The EPP should be developed as set out in US Federal Dam Safety Commission Guide (US-FDSC Guide[12]) and should be available, approved, and tested before the first impoundment of the reservoir (see Section 2.4.10 below).

### 2.4.3   Sedimentation

a.   Catchment Sedimentation load

Potential sediment yields should be assessed using a combination of analytical methods based on soil types, flow conditions, and comparison with similar dam sites. This assessment should then be verified through sampling undertaken at the dam sites at as wide a range of flow conditions and seasons as possible. An assessment of the proportion of silt deposited within the reservoir or passed downstream should be made using a relationship such as the Brune curve.[13] Where high sediment flows are anticipated, the risk of blockage and additional wear on structures, valves, and other assets should be considered in the design. Methods for assessing sediment yields are described in ICOLD Bulletin 115.[14]

b.   Sediment management

A whole life cycle cost approach to sediment management should be taken as described in the RESCON study,[15] effectively balancing the cost of construction of sediment management facilities such as silt traps, check weirs, or sediment flushing against the cost of future silt removal operations. Methods for managing sediment are also described in ICOLD Bulletin 115.

### 2.4.4   Geological, geotechnical, geotechnical, and geophysical conditions

This work typically includes reviews of the following:

• Review of geological mapping, plans, and cross sections showing all exploratory features and summarising drilling logs, geological interpretation, to include at least the dam, appurtenant structures, materials sources, and if available, the reservoir geology. Particular attention should be paid to geologic features which influence design considerations such as shear zones, fault open fractures or landslides, variability of formation consisting of liquefiable material, bedding planes, etc.

• review detailed exploration logs, including lithologic and physical conditions of material encountered, water test data, standard penetration or other resistance testing results, and frequency and type of samples obtained for laboratory testing

• review geophysical data

• review water level records of groundwater in the vicinity before and after the reservoir was filled

• review petrographic or chemical studies of foundation materials and natural construction materials

• review geological portions of all reports relevant to the site from preliminary reconnaissance studies to final as-built records

• review aerial photos of site and reservoir

• review published or unpublished regional geological studies that are relevant to the dam and reservoir setting

• examine the pertinent features of the aerial geology at the dam and the appurtenant sites, borrow and quarry sites, and, to the extent practicable, in the reservoir basin. Examine representative cores recovered from site explorations, particularly from zones indicated on the logs as being badly broken, weathered or highly pervious

- based on general geological setting, is this an acceptable site for the type of dam selected? Are bedding planes and joints particularly favourable or unfavourable to seepage, slope stability, foundation stability, acceptance of dam and reservoir loads and pressures, and sliding?
- was the effect of raised groundwater level on the stability of abutment and reservoir slopes considered?
- was potential chemical activity - reactivity of aggregate, quality of surface and groundwater, type of cement, etc. evaluated?
- was foundation susceptible to improvement by treatments such as pressure grounding, slurry grouting, blanket grouting, drainage, dental concrete, and deeper or more extensive excavation
- assess the adequacy of the overall exploration programme
- did geological information, gathered during construction, correlate with the information originally available to designers? If there are any significant differences, was the actual treatment of the geologic conditions adequate to compensate for changed conditions? If anticipated quantities or borrow and other materials were not obtained from primary sources, were the alternative sources sufficiently similar?
- was adequate geologic information available?
- identify all documents reviewed. List as references in the report being prepared

## 2.4.5  Seismicity

a.  Context

Historically, few dams have been significantly damaged by earthquakes (USDD, 2014[16]). On a worldwide basis, only about a dozen dams are known to have failed completely as the result of an earthquake. These dams were primarily tailings or hydraulic fill dams, or relatively old, small, earthfill embankments of perhaps inadequate design. About a dozen other embankment or concrete gravity dams of significant size have been severely damaged.

As dam infrastructure includes spillways, reservoirs, dams, tunnels, bridges, and other appurtenant structures, the seismic hazard assessment must be holistic and applied to the range of dam infrastructure as relevant.

b.  Data collection and considerations

Data on seismic and tectonic history of the region shall be collected and reviewed. When reviewing the earthquake data, considerations shall be given to the following:

- surface rupture
- Ground shaking and slumping
- Landsliding
- Liquefaction
- Seismic stability and settlements

Useful guidance is provided in the following ICOLD bulletins:

- ICOLD Bulletin 123, (2002). Seismic Design and Evaluation of Structures Appurtenant to Dams[17];
- ICOLD Bulletin 124, (2000). Reservoir Landslides: Investigation and Management[18];
- ICOLD Bulletin 113, (1999). Seismic Observation of Dams, Guidelines and Case Studies[19];

- ICOLD Bulletin 112, (1998). Neotectonics and Dams, Guidelines and Case Histories[20]

c.  Active/capable fault assessment

Damage to dams and their appurtenant structures may result from direct fault movements across the dam foundation, or more likely, from ground motion induced at the dam site by an earthquake located at some distance from the dam.

Active faults result in breakages (ruptures) at the ground surface and the generation of seismic ground shaking. Neotectonically active faults are those with surface breaks during Neogene-Quaternary while recent active faults have breaks in the Holocene Epoch (c. 11,000 years before present) (ICOLD Bulletin 112). Contemporaneous active faults have historic evidence of surface breaking. The basic property of an active fault with surface rupturing capability (or "capable" fault in terms used by the International Atomic Agency) is a reasonable probability of producing a surface break during the lifetime of the dam. Active faults if present can compromise dam safety and integrity and therefore must be investigated as part of assessing dam site feasibility and design.

d.  Liquefaction

A simplified desk study informed assessment of soil liquefaction and deformation should be undertaken initially to determine if seismic loadings control the design and to set the scope of any proposed size-specific liquefaction studies.

If further analysis of liquefaction hazard is required, it should include field investigations with appropriate geotechnical and geological sampling to assess the type and spatial distribution of foundation and embankment materials and the engineering properties of soil and rock.

Before beginning an evaluation of liquefaction potential appropriate ground motions must be established.

The propagation and duration of the ground motion through the foundation and embankment and liquefaction potential of the foundation and embankment soils should be assessed (Ref. FEMA-65, United States Bureau of Reclamation (USBR Guidance).[21]

Analysis to determine the static factor of safety immediately in after shaking is required.

Post-earthquake deformation analysis should be also undertaken.

Generalised settlement should be considered in addition to deformation of the embankment for slipping in response to earthquake shaking; the dam may settle in response to stresses developed in different soil layers.

e.  Review of Seismic Design Parameters

Typically, dams are to be assessed for the Safety Evaluation Earthquake (SEE) and an Operating Basis Earthquake (OBE). Reference for this assessment is ICOLD Bulletin 148.[22]

**SEE**

The Safety Evaluation Earthquake (SEE) is the maximum level of ground motion for which the dam should be designed or analysed. For dams whose failure would present a great social hazard, the SEE will normally be characterised by a level of motion equal to that expected at the dam site from the occurrence of a deterministically-evaluated maximum credible earthquake or of the probabilistically-evaluated earthquake ground motion with a very long return period, for example, 10,000 years.

Deterministically- evaluated earthquakes may be more appropriate in locations with relatively frequent earthquakes that occur on well-identified sources, for example near plate boundaries. It will be required at least that there is no uncontrolled release of water when the dam is subjected to the seismic load imposed by the SEE. Depending on the circumstances (e.g. the importance of the dam, the consequences of a dam failure) it is recommended to design safety-critical elements such as the bottom outlet and/or spillway gates for the SEE. Where there is not a great risk to human life the SEE may be chosen to have a lower return period depending on the consequences of dam failure. The above return periods are broadly in line with those being used for spillway design.

### OBE

The Operating Basis Earthquake (OBE) represents the level of ground motion at the dam site for which only minor damage is acceptable. The dam, appurtenant structures, and equipment should remain functional and damage should be easily repairable, from the occurrence of earthquake shaking not exceeding the OBE. In theory, the OBE can be determined from an economic risk analysis but this is not always practical or feasible. In many cases, it will be appropriate to choose a minimum return period of 145 years (i.e. a 50% probability of not being exceeded in 100 years). Since the consequences of exceeding the OBE are normally economic, it may be justified to use a more severe or less severe event for the OBE (i.e. longer or shorter recurrence period).

### Reservoir-Triggered Earthquake

The Reservoir-Triggered Earthquake (RTE) represents the maximum level of ground motion capable of being triggered at the dam site by the filling, drawdown, or the presence of the reservoir. There are a limited number of documented cases of reservoir-triggered earthquakes and a detailed study of such cases is recommended. General environmental features leading to RTE are detailed in the succeeding paragraphs. The ICOLD Bulletin 137[23] on Reservoirs and Seismicity provides the state of knowledge on reservoir-triggered seismicity.

The consideration of the RTE has been reported as generally linked to dams higher than about 100 m or to large reservoirs (capacity greater than about 500 Mm$^3$) and to new dams of smaller size located in tectonically sensitive areas. While there exist differences of technical opinion regarding the conditions which cause reservoir-triggered seismicity, it should be considered a credible event if the proposed reservoir contains active faults within its hydraulic regime and if the regional and local geology and seismic record within that area are judged to indicate the potential for reservoir-triggered seismicity. Even if all the faults within a reservoir are considered tectonically inactive, the possibility of reservoir-triggered seismicity should not be totally ruled out, if the local and regional geology and seismicity suggest that the area could be subject to reservoir-triggered seismicity. Depending on the dam location and prevailing seismotectonic conditions, the RTE may represent ground motion less than, equal to, or greater than the OBE ground motion. RTE ground motion should in no case be greater than the Safety Evaluation Earthquake ground motion and the faults considered capable of triggering seismicity should be taken into consideration during the seismic hazard evaluation. Still, the result might be the premature triggering of seismic events due to the impounding of the reservoir that would have occurred naturally at some longer time in the future. It is therefore justified in case of larger dams and storages located in seismically active regions and regions with high tectonic stresses to install a microseismic network and to monitor the seismicity prior to, during, and after impounding.

**Construction Earthquake**

For critical construction phases and temporary structures such as cofferdams, retaining structures, etc. it is also necessary to check the earthquake safety. The return period of such earthquakes depends on the type of structure, the duration of its use or the duration, and seismic vulnerability of the structure during critical construction stages and the consequences of its failure.

**Design Earthquake for Appurtenant Structures**

As a minimum, appurtenant structures should be designed in accordance with the applicable seismic code for buildings or other structures. Consequently, the site-specific design earthquake ground motion should have a return period equal to that specified in the seismic building codes, which is typically 475 years. For structures that are critical for dam safety, such as bottom outlet, spillway gates, control units and power supply, the design must be checked for the safety evaluation earthquake (SEE).

f.   Review of Seismic analysis

**Embankment Dams**

Stability of an embankment dam to earthquakes is affected either by loss of strength in the embankment or foundation material, or excessive permanent deformations (slumping, settlement, cracking of the embankment, and planar or rotational slope failures).

**Concrete Dams**

Safety concerns for a concrete dam to earthquake loading involve assessing the stability of the structure and ensuring the prevention of excessive cracking (overstressing) of the concrete.

## 2.4.6   Reservoir basin

a.   Rim stability/landslides

The potential for mass movement hazard needs to be assessed in relation to the reservoir basin. Risks of mass movements in the reservoir rim area must be evaluated and mitigation measures may be needed. The stability and structural integrity of the reservoir rim upstream of the structure must be evaluated for all potential loading conditions whether hydrologic, earthquake, or other hazards, man-made or natural.

b.   Basin Leakage

Basin leakage during impoundment occurs when infiltration of reservoir water is occurring through the surrounding and underlying rock and soils, which is problematic if seepage occurs beneath the dam retention structures and can cause other erosional issues for the reservoir area. The permeability, hydraulic conductivity, and porosity of the soils and rock supporting the reservoir volume should be evaluated. Erosion potential and corrosivity of the foundation soils and rock should be evaluated. Modelling of the infiltration rate of the reservoir should be performed. When there are high infiltration rates or the generation of high seepage pressures and velocities, a mitigation plan is needed. It should be considered in the design how the construction of an impermeable layer will be performed. Construction of impermeable layers and materials should consider construction-related damage and puncture as well as considerations for long-term maintenance and preservation.

## 2.4.7   Dam body

a.   Freeboard

The overflow capacity of the spillway must be sufficient to ensure that no overspilling of the dam occurs from the peak stillwater level in the design and safety check flood events. A freeboard should also be provided to ensure that effectively no wave overtopping occurs in a design event and only small wave overtopping flows occur in the safety check event, with such flows limited to that which can be safely tolerated without unsafe damage to the dam.

Wave run-up will be influenced by the dam geometry and materials, reservoir size and shape as well as wind speed and direction. Different combinations of reservoir stillwater level and wind velocity may occur. USBR Design Standard,[24] Chapter 6 provides a suitable approach for determining freeboard requirements for embankment dams. Appropriate wind speeds of different return periods and direction will need to be determined based on local data. Allowable wave overtopping rates are typically higher for concrete dams, though erosion of the ground at the dam toe and risk of dam instability still requires consideration.

An additional consideration for embankment dams is that the peak stillwater level in the reservoir should remain below any impermeable core or barrier within the dam with an allowance made for settlement and desiccation.

b.   Water tightness measures

Subject to foundation conditions watertight measures may be required. The type and robustness of the watertight measures should be carefully considered to ensure the dam performs the design intent, including during extreme events, for example, seismic and flooding. The form of seepage control or water barrier within the dam will depend on the choice of type of dam as described above. As well as the risk of seepage or leakage, and therefore internal erosion, through the dam body seepage through the dam foundation should also be considered. In a large number of cases, a seepage cut-off below the dam is required which, if shallow, could include constructing a low permeability trench such as from clay. Often deeper cut-offs are required to provide a continuous barrier against seepage between the dam body and a relatively impermeable geological strata below the dam. Examples include cement-bentonite or grout trenches, cut-off walls such as diaphragms and sheet piling or curtain grouting. Consolidation grouting below the dam may also be required.

c.   Dam Stability

   i.   Embankment Dams

Guidance on design of embankment dams including slope stability is provided in USBR DS13 Embankment dams and Design of Small dams.[25]

All global stability analysis for dams should consider, at least, the four key critical modes, which include, but are not limited to:

a.  End of Construction – Drained and undrained analysis is performed, undrained properties for embankment materials for when wet cohesive materials are used for construction.

b.  Rapid Drawdown – Undrained analysis is undertaken where the water level is high for a prolonged period and then lowers quickly, which represents conditions right after an extreme water event. Particularly applicable to upstream face of the dam and the reservoir rim.

c.  Steady State Seepage – Evaluation of steady-state water level conditions, which represents the phreatic surface through the embankment during extreme water event. Submerged unit weights of soils beneath the phreatic surface should be used. Drained and undrained mechanical properties should both be used in the analysis.

d.  Earthquake – Evaluation of embankment stability when a seismic event occurs. For conditions where liquefiable soils are identified beneath the dam or appurtenant structures, the factor of safety against liquefaction should be assessed and the levels of liquefaction induced settlements. Seismic analysis will be carried out at full supply level for an OBE and SEE

Static and dynamic analysis of the dams should include:

a.  2D finite element analysis to evaluate the seepage and hydraulic pressures developed in the foundation materials, in dam structure (if earth fill), and the hydrostatic pressures built up against all retention structures.

b.  Static 3D finite element analysis to evaluate total and effective stresses developed in the foundation materials and in the dam structural elements.

c.  Jointing and faulting of the basement rock should be explicitly modelled.

d.  Perform time-domain dynamic 2D finite element analysis incorporating drained, undrained, and coupled. Determine areas of transient and permanent deformation.

e.  Analysis of the soil-structure interaction for structures interfacing with foundation soils or rock should be performed.

f.  The modelling and analysis should determine the stresses and deformations occurring in the structural elements of the dam and appurtenant structures.

Load combinations and factors of safety should be adopted such as those in USBR DS13 Embankment Dams and Design of Small Dams. The following loads should be considered as a minimum:

1.  Dead Loads – of the dam or appurtenant structures

2.  Water Load (Reservoir) – the peak reservoir water level shall be determined following the hydraulic design of the spillway. Associated pore water pressures shall be considered. Hydrodynamic water pressures shall be applied in seismic conditions

3.  Water Load (Tailwater) – it is anticipated that a hydraulic jump stilling basin shall be provided to dissipate the energy of the flow in the spillway. The action of the hydraulic jump will move the tailwater downstream of the dam and so may remove or reduce the restoring downstream tailwater.

4.  Silt – confirmation is required of the anticipated siltation of the reservoir before an assessment of this load can begin.

5. Earth loading– weight and earth pressures
6. Ice – if applicable
7. Earthquake – two scenarios shall be considered: the OBE and SEE, as discussed above
8. Surcharge such as maintenance vehicles or mobile equipment such as that which may be required for emergency response

ii. Concrete Dams

Design considerations for the stability of a concrete dam are similar to those of an embankment dam with the obvious exception that the slopes of the dam are not subject to geotechnical analysis, the slope stability of other earthworks and the reservoir rim however remains an important consideration.

1. Uplift – whilst seepage through the dam body is less of a concern than for an embankment dam, seepage through the foundation and uplift pressures are generated by the reservoir water load and distributed linearly across the dam footprint, as the worst-case scenario. The uplift will be a full reservoir load at the upstream toe and tailwater level at the downstream toe.
2. Global Stability – the key failure mechanisms for concrete dams are overturning, sliding at the dam/foundation interface, and bearing capacity failures.
3. Stability of Blocks – the interface of individual lifts, blocks, or elements needs to be considered with the stability of each portion of the structure being ensured
4. Temperature – stresses arising from thermal effects as well as the location and detailing of joints require careful consideration

The loads to be considered are typically similar to those experienced by embankment dams as described above, design guidance for gravity dams (for example) is provided in the aptly titled USBR "Design of Gravity Dams" and similar guidance exists for other common concrete dam types.

d. Seepage

Seepage analysis is critical for dams, though some seepage is inevitable, ensuring that seepage through the embankment and its foundations (or the foundations of concrete dams) is limited to safe levels. Excessively high seepage flows may cause internal erosion, washing out fine material leading to dam breaches. Excessive seepage can also cause flooding or other issues around the reservoir rim.

For embankment dams, a combination of measures is typically implemented to reduce the risk of such failures, impermeable barriers like dam cores and below dam cut-offs reduce seepage flows, filter layers are typically included downstream of fine-grained materials such as dam cores to prevent the loss of fine material.

The grading of foundation materials and the seepage path length will depend on whether the foundation is vulnerable to internal erosion. Other protective measures include drainage blankets and rock toes for embankments.

Guidance on internal erosion is provided in ICOLD bulletin 164.[26]

e. Drainage

Embankment dams typically include filter drains along the toe of the dam to collect seepage flows and transport them to a monitoring point such as a v-notch weir. Some embankment dams also include chimney drains and drainage blankets to safely collect the seepage. Collection and monitoring of these flows are important to detect trends in flow, an unexplained increase in which may be a symptom of internal erosion or another issue within the dam.

Concrete dams and some spillways may include drains and drainage galleries to collect seepage from under the structures to reduce the risk of erosion and uplift pressures on the structure.

Spillway structures often also include under drainage, particularly at joints, this is described elsewhere.

Guidance on the design of drainage is included in the USBR Design of Small dams.

### 2.4.8   Overflow spillway

Each dam will require one or more overflow spillways to safely convey the design and safety check floods. Different requirements would apply in each case for example with greater freeboard requirements applied to the design flood than the safety check. The overarching definition of design and safety check floods described earlier means that a spillway must comfortably pass the design flood with the spillway and dam remaining safe with an additional margin of safety remaining available. In a safety check flood, some damage may occur to the spillway and dam but this must not threaten the safety of the dam.

Numerous forms of spillway may be possible to suit the form of dam and hydraulic requirements. The form and design of spillways are discussed in ICOLD Bulletins 58[27] and 172.[28] The location of spillways is also important with the resulting risk of a failure of the dam being considered, both due to design flow exceedance and due to defects or structural failures with the structure or its interface with the dam or ground. Risks such as seepage from or around the spillway or structural failure should be considered. The preference is therefore typically for spillways located off the dam or at its abutment, though this is not always achievable.

The design of the spillway should typically ignore any flow through outlet works where smaller pipes, valves, and screens are likely to be blocked, the spillway therefore carries the full design and safety check outflows.

Spillway options include gated or uncontrolled spillways. Gated spillways require expensive mechanical and electrical components and require additional specialist maintenance and operation. Determining a safe operating plan also requires careful consideration, balancing operator response times, rate of flood rise, and balancing reservoir safety with downstream flood risk. Passive spillways, without gates or other operating systems, are therefore preferred. Where a gated system is considered necessary the whole life cost-benefit and technical necessity of the system should be demonstrated.

The crest level of the spillway will be set based on the required storage capacity within the reservoir such that the floods up to the design capacity of the downstream river defences are contained within the reservoir without spilling, or that it is demonstrated that this spill does not pose a flood risk downstream. In events exceeding the design standard of the river defences the reservoir spillway would discharge to ensure the safety of the dam.

For high hazard dams, it is expected that the performance of the spillway is demonstrated through a sufficiently detailed physical or Computational Fluid Dynamics (CFD) model. For lower hazard dams a model is also likely to be required unless the designer can demonstrate to the client's satisfaction that the design is sufficiently simple and of standard design that previous research or model testing of similar structures can be relied upon to demonstrate satisfactory performance.

In all cases, it will be necessary to demonstrate the discharge performance of the spillway from zero to the safety check flood or greater and to plot a rating curve of this performance. This rating curve should inform the routing analysis of floods through the reservoir described elsewhere, with the spillway design safely passing the outflow. The approach channel must be designed to not restrict the performance of the spillway and to provide approach velocities low enough to minimise scour risk. The risk of debris blockage and impact damage to the structure should be considered in the design with upstream booms or log catchers considered necessary.

It will be necessary to explore the flow velocity and depth along the length of the chute. Phenomena such as air bulking and cavitation will need consideration along with an assessment of uplift/plucking pressures, hydrostatic and hydrodynamic effects. The spillway chutes must be of a robust, non-erodible material and it is anticipated that the spillways will be predominantly reinforced concrete. Spillways should be straight and of constant gradient or with gradual changes in direction and gradient where these changes are carefully analysed with consideration of the risk of super-elevation and cross-waves.

At the base of each spillway a terminal structure such as a stilling basin to ensure that energy from the chute flow is dissipated, and flows discharged at safe velocities with the outlet protected from scour. There are a wide variety of designs including flip buckets and baffled or unbaffled basins. The approach and different forms of terminal structures are described in USBR "Design of Small Dams" and "Hydraulic Design of Stilling Basins and Energy Dissipators". These guides provide design guidance on standard forms of structure based on model testing, other forms and variations are possible but may require additional model verification. In order to correctly design these terminal structures and determine the outlet conditions, it is necessary to determine the flow conditions in the downstream channel over a range of flows, developing a tailwater rating curve. This would typically be developed from numerical modelling of the downstream channel.

Freeboard within the spillway should be provided using the approach described in USBR Design of Small Dams for both the approach, chute, and stilling basin.

The structural detailing of spillways is important both to safely carry the required floods but also so as to manage risks with the interface with the dam or foundation. In particular, the structure should be detailed to prevent seepage between the spillway and its foundation and to ensure the watertightness of the structure. Movement joints are a particular area requiring careful attention to ensure sufficient flexibility again movement and watertightness. Guidance on detailing is provided in USBR guides "Design of Small Dams" and "DS14-3 Appurtenant Structures for dams".[29] These references also provide guidance on the appropriate structural design considerations such as load cases.

### 2.4.9   Outlet works

a.   Pass-forward control structure

Outlet works shall be designed to release the intended pass-forward flow, as such the flow from the outlet works should be to suit the requirements of the river flood alleviation scheme measures. The dam design therefore requires careful coordination with the river flood alleviation works to ensure the outflow suits the river flood scheme design and that there is sufficient storage within the reservoir to store the difference between the design storm inflow and maximum pass forward flow.

Wherever possible the design of outlet works should provide flexibility to allow the pass forward flow to be easily altered in future, for example by incorporating a replaceable orifice plate or flow control valve. The design of the outlet works should consider the safe maintenance, inspection, and operation access to culverts, pipes, and flow controls.

Preference shall be made for passive flow control methods such as orifice plates and vortex flow controls (e.g. hydrobrake valves) over actively controlled mechanical gates and valves due to the lower maintenance and operation requirements and greater reliability. However, where they offer significant benefits actively control methods may be necessary providing operation and maintenance considerations can be adequately addressed. The ideal flow control would permit flows smaller than the downstream design pass forward flow to exit the reservoir without impoundment in the reservoir so as to preserve storage space for the peak of a flood but prevent any flows larger than the design from leaving the reservoir. A cost-benefit assessment should be made between more elaborate, better performing, but potentially more expensive valves and control structures which may minimise the size and therefore cost of the dam versus simpler but less efficient valves requiring greater reservoir storage and larger dams. Rating curves should be developed for the performance of the proposed options and tested through 1D numerical hydraulic routing models to verify the outflow and peak reservoir water level in the relevant storm events over a range of storm durations.

Outlet works should be equipped with appropriately sized and arranged debris and security screens to reduce the risk of blockage and unauthorised access. Consideration must be given to the access, maintenance, and clearance of these screens.

All penetrations of the dam such as the outlet works will require careful detailing to reduce the risk of seepage at the interface between the structure and dam.

b.   Emergency drawdown facilities

Emergency drawdown facilities are provided to allow the reduction in the water level of a reservoir if an outlet is obstructed, or a more rapid drawdown is required such as in the event of a dam emergency. A bottom outlet, emergency drawdown facility should be provided for each dam, this may be combined with or separate from sluices for sediment flushing if the geometry suits. Larger dams may require additional drawdown capacity and so typically require separate and additional facilities. Design of outlets for sediment flushing is detailed in ICOLD Bulletin 115.

The emergency drawdown facilities should be separate from the main outlet such that a blockage of one does not affect the other. The emergency drawdown facilities should be sized at an appropriate amount of reduction in reservoir water level within a set time to reduce the load on the dam and to reduce the hazard to the downstream population. This drawdown should be achieved whilst there is a certain inflow into the reservoir. These parameters should be based on the hazard classification of the dam and international guidance such as USBR ACER Technical Memorandum No. 3.

Inappropriate or over-rapid drawdown may lead to slope instability in an (embankment) dam or the reservoir rim. In addition, high discharges may lead to scouring of the inlet, or outlet tailrace. It is therefore important that these risks are considered not only in the design of the dam and drawdown facilities but also in the operation procedures for and maintenance of these facilities (refer to Operation and Maintenance 2.4.10(b)). All penetrations of the dam such as the drawdown facilities will require careful detailing to reduce the risk of seepage at the interface between the structure and dam.

c.   Supply draw-off

Supply draw-off is usually achieved through an inlet with a pipe and valve system, a specific supply draw-off is only required if the dam is designed to achieve multiple purposes such as for irrigation or water supply. Guidance for the design of supply draw-offs can be found in the USBR Design of small dams. It is important that the ongoing operation and maintenance of this system and valves are considered in the design. All penetrations of the dam such as the outlet works will require careful detailing to reduce the risk of seepage at the interface between the structure and dam.

## 2.4.10   Review of instrumentation, operation and maintenance and emergency preparedness plans

a.   Instrumentation Plan

ICOLD Bulletin 61[30] describes four important purposes for instrumentation of a dam:

a.   To indicate the evolution of conditions during construction so that the validity of certain criteria can be confirmed before the project is commissioned
b.   To indicate the evolution of conditions during and following the first filling of the reservoir so that performance can be evaluated with respect to that assumed in the development of the criteria
c.   To indicate the evolution of undesirable conditions throughout the life of the dam so that remedial measures may be undertaken to avoid deterioration of its security
d.   To record the experience for the improvement of the design of future dams

A programme of instrument readings, combined with inspections and periodic evaluation reports throughout the life of the dam, will serve to confirm that the security criteria of the design remain valid or that steps should be taken to restore the dam and its foundation to acceptable levels of security (ICOLD 61).

ICOLD Bulletins 60,[31] 68,[32] 87,[33] 118,[34] 138,[35] 158[36] and 180[37] further describe the objectives and forms of monitoring for dams as part of an overall package of surveillance. Consideration should also be given to automating elements of the instrumentation and monitoring, particularly in remote sites as described in

ICOLD Bulletin 118, though this should be seen as a supplement to visual inspections rather than a replacement.

The design of the instrumentation for each dam will differ depending on the form of the dam, its design, and site conditions. Typically monitoring should include:

- Reservoir water-level monitoring and recording
- Seepage and Leakage Monitoring – such as with v-notch weirs for flow monitoring and inspection of flow
- Water pressure monitoring in (embankment) dams and in foundations (all dams)
- Deformation monitoring such as crest settlement and displacement or for deformation of slopes
- Seismic monitoring/strong motion detectors

Surveillance and monitoring are important through the life cycle of the dam; however, additional significance should be taken during the first filling of the reservoir. Regular surveillance observations of the dam and reservoir should be undertaken by suitable trained personnel. The frequency of the visits can be related to the hazard posed by the dam.

It is usual international practise to employ the four eyes principle, with the first pair of eyes belonging to the owner or dam operator and the second pair to an independent body, for example supervisory authority or panels of experts (ICOLD 167 Regulation of Dam Safety:

An overview of current practice worldwide[38]). ICOLD 180 Dam Surveillance – lessons learnt from case histories provides worldwide examples of surveillance and why it is important. A monitoring regime should be established to take measurements or readings relevant to the type of dam. Analysing the measurable parameters can suggest the acceptable performance of the dam or indicating deviation from expected behaviours, prompting more detailed surveillance or monitoring and/or interventions to protect the safety of the dam.

b.  Operation and Maintenance Manual

The USBR guide for the preparation of Operation and Maintenance (O&M) manual,[39] and procedures for dams could be used as a framework to develop a site-specific document. There should be one primary controlled document with the complete, accurate, and current operating instructions for each storage reservoir and its related structures. The purpose is to ensure adherence to approved operating procedures over long periods of time and during changes in operating personnel.

The O&M manual shall also address procedure for sediment management, as discussed in Section 2.4.3.

The document extends beyond the everyday operations to procedures required in an emergency, for example, where the dam is at risk of an uncontrolled release of the reservoir water.

Alternative guidance is provided in ICOLD Bulletin 154.[40]

c.  Emergency Preparedness Plan

An outlined EPP is typically prepared prior to construction and agreed with relevant stakeholders. A typical content of an EPP, based on the recommendations by the US Federal Dam Safety Commission,[41] is given below.

1.  Introduction and objective of the EPP
2.  Description of the scheme

3. Emergency detection/failure modes
4. Notification Flowchart
5. Alarm Levels and Organisation of Warning
6. General Responsibilities under the EPP
7. Notification Procedures and response matrix
8. Early Warning System (EWS)
9. Mitigation Activities
10. Organisation of evacuation
11. Approval, maintenance, exercise and correction of EPP
12. Appendix (dam break analysis, inundation maps, etc)

The notification flowchart, Section 4 of the EPP, shall list all stakeholders and shall show clearly, for each level of emergency, who is to be notified, and who is responsible for notifying which owner representative(s) and/or public official(s), and in what priority. The notification flowchart shall include individual names and position titles, office and home telephone numbers, and alternate contacts and means of communication.

The alarm levels and organisation of warning (Section 5 of the EPP) shall include the following alarm levels:

• Minor Deficiencies – Level 1: mobilise personnel and equipment to deal with minor deficiencies
• Serious Deficiencies – Level 2 – Local public officials have to be informed to be prepared for action if the situation gets worse
• Very Serious Deficiencies – Level 3 – as for Level 2
• Alarming Deficiencies – Level 4 – Special Mobile Force, police and local public to be alarmed; radio broadcasting to be interrupted to warn the population; the population to be evacuated

Section 6 describes general responsibilities under the EPP. It typically describes:

• operators' duties in implementing the EPP
• the person(s) authorised to notify local officials should be predetermined & clearly set forth in the EPP
• gives pointers on how to communicate the emergency situation to those who need to be contacted
• describes predetermined remedial action to delay or mitigate the severity of failure
• EPP should be co-ordinated with high enough levels of management to ensure full awareness of capabilities
• describes specific actions operators are to take after implementing the EPP

The EWS shall include the following:

• Dam monitoring instruments
• Warning equipment (sirens, etc)
• Computer/printer
• Communication system

## NOTES

1  ICOLD bulletin 157 – Small Dams_ Design, Surveillance and Rehabilitation, 2010.
2  ICOLD bulletin 72 – Selecting Seismic Parameters for Large Dams, 1989.
3  United States Federal Energy Regulatory Commission (FERC), Engineering Guidelines for the Evaluation of Hydropower Projects, 2003.
4  USBR guide on Safety Evaluation of Existing Dams.
5  ICOLD Bulletin 170 – Flood Evaluation and Dam Safety.
6  FD2628 Impact of Climate Change on Dams & Reservoirs: Final Guidance Report May 2013, the Department of Environment, Food and Rural Affairs UK.
7  Climate-resilient Infrastructure OECD ENVIRONMENT POLICY PAPER NO. 14, 2018.
8  Climate change impacts on dam safety: Javier Fluixá-Sanmartín et al., May 2020.
9  ICOLD Bulletin 82 – Selection of Design Flood.
10  ICOLD Bulletin 125 – Dams and Flood Case Studies.
11  ICOLD bulletin 111 – Dam Break Flood Analysis.
12  US Federal Dam Safety Commission Guide (US-FDSC Guide).
13  Reservoir Sedimentation Handbook, Design and Management of Dams, Reservoirs, and Watersheds for Sustainable Use, Gregory L. Morris, Jiahua Fan, 1998.
14  ICOLD Bulletin 115 – Dealing with Reservoir Sedimentation.
15  Reservoir Conservation Volume 1: "The RESCON Approach" (World Bank, June 2003).
16  USSD Committee on Earthquakes – Observed Performance of Dams During Earthquakes Volume III, 2014.
17  ICOLD Bulletin 123 – Seismic Design and Evaluation of Structures Appurtenant to Dams.
18  ICOLD Bulletin 124 – Reservoir Landslides: Investigation and Management.
19  ICOLD Bulletin 113 – Seismic Observation of Dams, Guidelines and Case Studies.
20  ICOLD Bulletin 112 – Neotectonics and Dams, Guidelines and Case Histories.
21  FEMA-65, United States Bureau of Reclamation (USBR Guidance).
22  ICOLD Bulletin 148 – Selecting Seismic Design Parameters for Large Dams-Guidelines.
23  ICOLD Bulletin 137 – Reservoirs and Seismicity.
24  USBR Design of Small dams.
25  USBR DS13 Embankment dams.
26  ICOLD Bulletin 164 – Internal Erosion on existing dams.
27  ICOLD Bulletin 58 – Spillways for dams – *State of the art.*
28  ICOLD Bulletin 172 – Technical advancement in spillway design.
29  USBR DS14-3 Appurtenant Structures for dams.
30  ICOLD Bulletin 61 – Dam design criteria – Philosophy of choice – *The philosophy of their selection.*
31  ICOLD bulletin 60 – Dam Monitoring- General Considerations.
32  ICOLD Bulletin 68 – Monitoring of dams and their foundations – state of the art.
33  ICOLD Bulletin 87 – Improvement of existing dam monitoring – Recommendations and case histories.
34  ICOLD Bulletin 118 Automated dam monitoring systems – Guidelines and case histories.
35  ICOLD Bulletin 138 – General approach to Dam Surveillance *Basic elements in a "dam safety" process.*
36  ICOLD Bulletin 158 – Dam surveillance guide.
37  ICOLD Bulletin 180 – Dam Surveillance – Lessons learnt from case histories.
38  ICOLD Bulletin 167 Regulation of Dam Safety.
39  United States Bureau of Reclamation (USBR) guide for preparation of Operation and Maintenance (O&M) manual.
40  ICOLD Bulletin 154 – Dam safety management: Operational phase of the dam life cycle.
41  US Federal Dam Safety Commission – Emergency Preparedness Planning.
42  US Army Corps of Engineers (2013), *Safety of Dams – Policy and Procedures.* ER 1110-2-1156.

Chapter 3

# Case studies

## Eurasia/Western Asia – Armenia

### 3.1  ABOUT THE COUNTRY

Armenia is situated in Eurasia/Western Asia, bordering Georgia, Azerbaijan, Iran and Turkey, see Figure 3.1 below. It is a landlocked country.

The terrain in Armenia is very mountainous with many rivers and forests. The land rises to its highest point of 4,090 m above sea level at Mount Ararat, in the south-west. The country is the tenth highest in the world by its average elevation and it has 85.9% of the mountainous area, more than Switzerland and Nepal. It is also home to Lake Sevan in the highlands which is the second largest lake in the world relative to its altitude of 1,900 m above sea level.

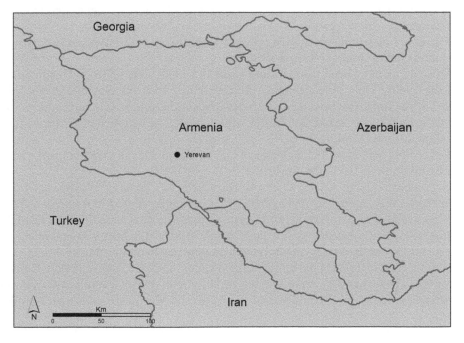

*Figure 3.1* Map of Armenia.

DOI: 10.1201/9780429320453-3

The modern Republic of Armenia became independent in 1991 during the *dissolution of the Soviet Union*.

**No. of people** – 3 million
**Area** – 29,743 km$^2$
**Population Density** – 101/km$^2$
**Economy:**
Gross Domestic Product (GDP) Nominal – (Total) $13.44 billion
Gross National Income[i] (GNI) Nominal – (per capita) $4,680
**Climate** – The climate in Armenia is continental highland, with dry, hot summers lasting from June to September, and the temperature ranges from 22°C to 36°C. The winters are cold with a lot of snow, the temperatures ranging between −10°C and −5°C.

The complex topography has a direct and indirect influence on the climate and especially the precipitation regime of the country. The mean annual precipitation varies widely from over 1,000 mm on many of the mountain ranges to about 400–700 mm over much of the country and to below 300 mm near the River Arakhs. The mean precipitation is strongly related to elevation, and a significant portion of the winter precipitation falls as snow; however, this is not as large a proportion of the total annual precipitation as might be expected, as the seasonal distribution is not dominated by winter precipitation.

## 3.2  ABOUT THE DAMS

**Number of dams:** 92 (83 irrigation, 6 hydropower, 3 others)
Most of the dams in Armenia are used for irrigation; they help to reduce the country's dependency on food imports and provide both security and socio-economic benefits.

The hydropower dams meet around a quarter of the demand for electricity. The rest of the energy is provided by nuclear reactors and fossil fuels, generated from thermal stations.

Armenia's dams include a wide range of height from 1.5 to 83 m and are structurally embankments or concrete gravity dams.

Most dams have been in operation since the 1960s and 1970s, with some in use since the 1940s. They were designed in accordance with the Soviet Design Norms (SNIPs) at the time. Due to a lack of proper inspections and maintenance, three complete dam failures occurred in 1974, 1979 and 1994. The number of fatalities due to these dam failures is unknown.

Since these accidents, the Government of Armenia put the safety improvement of the existing dams as the highest priority.

Over a period from 2000 to 2009 two Dam Safety Projects (DSPs) financed by the World Bank (WB), namely DSP I and DSP II[1] were implemented in Armenia. The two DSPs involved safety assessments of 87 existing dams and, as a result of the two DSPs, 74 dams were rehabilitated since; 20 dams were considered to pose extreme safety risks (Risk Class IV, see Section 2.2.3) and required immediate rehabilitation.

Among the portfolio of dams were also two uncompleted dams, namely Kaps (see Section 3.3.2 below) and Marmarik[2,3] which have been posing a significant risk to the downstream population. Rehabilitation works for the Marmarik dam were completed

---

i As per the World Bank Classification.

in 2012 under DSP II. The design of safety rehabilitation works for the Kaps dam has now been completed under a different project and it is expected that the construction works will take place in 2022.

Also, included in the DSPs was the safety assessment of the Bartsrouni dam, which was constructed on a large, ancient landslide, while recent landslide activities have been demonstrated by numerous scarps. The dam has already been partially destroyed by landslide movements and it is anticipated that future movements will continue to damage the dam.

The main safety aspects identified and addressed in the DSP II are described below in Section 3.3. Some specific safety aspects of Marmarik Dam rehabilitation are described in Section 3.4.

**Law and regulations**

There are no specific laws and regulations related to dams' safety. Standards used for dam design are largely based on the SNIPs. Emergency Preparedness Plans are normally produced for high hazard dams and are kept by the Ministry of Emergency Situations.

**ICOLD membership**

Armenia is a member of ICOLD.

## 3.3   DAM SAFETY PROJECTS II

### 3.3.1   Scope of the project

The DSP II, which was undertaken in 2003–2004 involved a safety assessment of 66 existing dams. The dams were classified as follows:

- Large dams (14 dams, 15–85 m high)
- Small dams (33 dams, 1.5–15 m high)
- Artificial lakes (17 dams, 0–5 m high)
- Partially constructed dams (2 dams, 14 and 51 m high)

Six of the large dams are for hydropower, while the other dams are for irrigation or multi-purpose.

The following work was undertaken under the DSP II:

- Dam inspection
- Field Investigations – Site investigations, Topographic survey, microseismic survey
- Studies – Hydrology, Flood routing, Dam break, Stability analysis, Seismic hazard assessment, seismic analysis
- Rehabilitation works preliminary design and costing
- Risk assessment
- Dam safety plans (Operation & Maintenance, instrumentation, early warning systems and Emergency Preparedness Plan)

A more detailed description of field investigations and studies is given in Sections 3.3.2 and 3.3.3 below.

Typical rehabilitation works proposed are described in Sections 3.3.4 and 3.3.5.

The Portfolio Risk Assessment carried out and the Dam Safety Plans are explained in Sections 3.3.6 and 3.3.7 respectively.

## 3.3.2   Field investigations

Only limited information existed regarding the construction of each dam, and typically the information available was design data rather than construction records. For many of the smaller dams, no records were found at all. Thus, although an archive search was carried out, it was necessary to carry out field investigations on most of the reservoirs including topographic survey and mapping, and geotechnical site investigations to obtain data for safety assessments.

Intrusive investigations were carried out at 32 dams and comprised:

- drilling of 4,000 m of boreholes, logging, in-situ testing, sampling
- excavation of trial pits, logging, sampling
- laboratory testing

Geophysical investigation was also undertaken and it comprised:

- Seismic refraction
- Electrical resistivity

## 3.3.3   Studies

a.  Hydrology

Two methods have been used to analyse the flood inflows into the reservoirs. The first, the SNIP method, is based on standard Russian techniques and is generally used in Armenia. The second, a statistical method, or the Regional Method, using all annual maxima flow data recorded in Armenia, has been used worldwide to check more particular methods.

The Regional Method depends on the analysis of annual maximum floods measured at the gauging stations within a region. It derives from the approach developed during the investigation of floods in the British Isles[4] and it has since been applied to many different regions of the world.

The method is based on the combination of records from a number of stations to give a representative record of a long duration. The maximum flows in each year of record at each station are reduced to non-dimensional form by dividing the annual flood by the mean annual flood estimated from the whole period of record at the station. Forty years' records on 102 gauging stations have been used for Armenia for the application of the Regional Method.

*Results of SNIP and Regional Method*

The SNIP results found higher estimates of peak flow for smaller catchments of up to (100 km$^2$); and for very large catchments (100,000 km$^2$), the regional methods gave a greater estimate, with a reasonable agreement between the two methods (Figure 3.2).

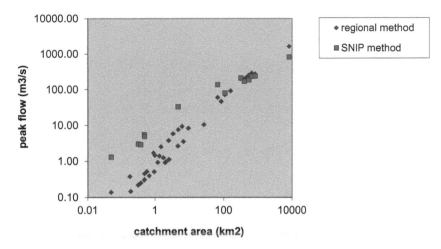

*Figure 3.2* **Peak flows vs catchment areas.**

b. Flood Routing

Flood routing studies were carried out making use of either inflow hydrographs based on SNIP hydrology and SNIP rules for the return periods to be considered, or inflow hydrographs based on Regional Method hydrology and ICOLD recommendations for return periods where this gave larger floods. In summary:

• Large Dams – for most of the large dams the design flood was 1 in 1,000 years and the safety check flood 1 in 10,000 years, except for the ones where a large population downstream was at risk, for which the design flood was 1 in 10,000;

• Small Dams – design flood was 1 in 100 years and the safety check flood was 1 in 1,000 years;

• Artificial Lakes – design flood was 1 in 20 years and the safety check flood was 1 in 100 years.

c. Seismic Studies

Seismicity in Armenia is high as the country is in the Caucasus, which lies between the Black Sea and the Caspian Sea, within a broad zone of deformation, that forms a part of the Alpine–Himalayan collision belt. The motion of the Arabian plate northward relative to the Eurasian plate dominates the present-day tectonics of the area.

The initiation of the continent-collision caused the folding and thrusting of the Greater Caucasus upwards and they are now the highest mountains in the western segment of the Alpine-Himalayan belt, see Figure 3.3. The Eurasian and Arabian plates converge at 28 mm/year along longitude 26°N near the Caucasus. Subduction of continental crust is avoided by lateral extrusion of the Turkish block, Azerbaijanian block and northwestern Iranian block and by crustal thickening through underthrusting of accreted terrains.

The seismicity of the area reflects the general tectonics of the region. Seismic activity increase could be noted in the region, connected with destructive events such as Spitak earthquake in Armenia, 1988 (M = 6.8), Racha earthquake in Georgia, 1991

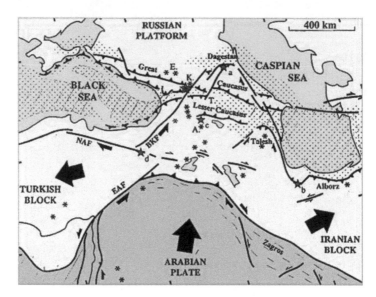

*Figure 3.3* Schematic map of main tectonic features of the Caucasus and adjacent territories.[5]

(M = 7.0), Barisakho earthquake, Georgia 1992 (M = 6.5), Erzinjan earthquake, Turkey 1993 (M = 7.2) etc.

The largest devastating earthquake that hit Armenia in 1988, the Spitak earthquake, had a surface wave magnitude of 6.8 and a maximum MSK intensity of X. The earthquake left 38,000 people dead and 31,000–130,000 injured. This earthquake had a big impact on the Kaps dam, located some 40 km away from the epicenter of the Spitak earthquake. The dam, originally planned to be 80m high, was at the time under construction, where about 20 m high embankment and diversion works were already built. The construction of the dam was since abandoned and the dam remains unfinished to date (see Section 3.2).

Figure 3.4 below presents the current Seismic Map of Armenia which delineates four Seismic Zones with associated seismic design coefficients as follows:

- Zone IV – seismic horizontal acceleration 0.5 g
- Zone III – seismic horizontal acceleration 0.4 g
- Zone II – seismic horizontal acceleration 0.3 g
- Zone I – seismic horizontal acceleration 0.2 g

It shall be noted that the above seismic accelerations have been defined for a return period of an earthquake of 1 in 475 years, which is not adequate when dams are assessed to SEE.

According to ICOLD bulletin 72 (Endnote 2 in Chapter 2), when assessing the safety of large dams, the dams shall be checked for OBE and MDE[ii] earthquakes.

---

ii  At the time the safety assessment was undertake the maximum design earthquake was the MDE, as per ICOLD Bulletin 72. Since then, the SSE earthquake has been introduced in Bulletin 148.

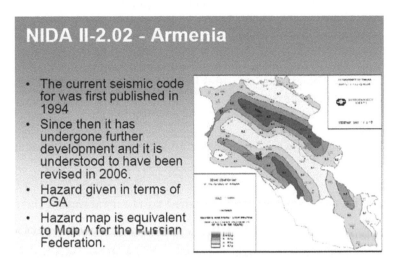

*Figure 3.4* **Seismic zones in Armenia.**

Seismic design parameters for assessing the seismic performance of dams have been selected based on:

*   Seismic Design Standards of the Republic of Armenia (SDSRA)[6] – for all dams
*   Site-Specific Seismic Hazard Assessment (SSSHA) – large dams. The SSSHA defined seismic input parameters for OBE and SEE/MDE earthquakes for large dams.

For the large dams, the seismic accelerations for the MDE were larger than the ones given for Zone IV.

### 3.3.4   Typical deficiencies and structural rehabilitation works proposed for existing dams in operation

A summary of defects that were identified relating to construction, design, maintenance, and operation are as follows:

*   Deliberate blockage of the spillway to increase storage.
*   Inadequate spillway capacity/freeboard.
*   Structural repairs required to spillway or outlets.
*   Damaged or deficient riprap or wave protection.
*   Outlet valve refurbishment required.
*   Slope stability inadequate.
*   Leakage through embankment.
*   Leakage through reservoir floor.
*   Unsafe access to equipment.

- Refurbishment required to hydromechanical equipment.
- Global instability

Based on the results of the assessment of defects, remedial works were recommended, and detailed designs were prepared by Armenian consultants.

There were a few emergency measures that were recommended to be immediately addressed after the inspections.

The Bartsrouni dam was proposed to be decommissioned, as it was built on an active landslide.

Pending remedial works, it was advised to maintain the reservoirs at lower water levels.

### 3.3.5   Specific dam safety issues related to the Marmarik dam

a.  Description

Marmarik Dam, constructed on the Marmarik River, was designed as a 64 m-high embankment dam with a clay core and compacted gravelly fill shoulders[3] (Endnote 3). A total volume of the embankment fill was 5 MCM. The dam was to impound a reservoir of 36.9 MCM at a full supply level (FSL) at elevation of 1,911 masl. The crest level was designed to be at elevation 1,914 masl, resulting in a freeboard of 3 m. The project was to irrigate 900 ha of farmland and supply water to the future aluminium mining industry, a cement factory and two thermal power plants.

The central part of the dam was founded mainly on granular river alluvium and the abutments were founded on a thick layer of cohesive, colluvial deposits. Foundation anti-seepage measures comprise a bored secant pile cut-off constructed up to 30 m deep through the central part of the alluvial foundation and a grout curtain through the colluvial foundation at the abutments. The upstream face was protected against the wave action with cast in place concrete slabs.

In November 1974, the embankment was completed to its full height. Twenty days later, only a few days before a scheduled closure of the diversion tunnel, the dam crest slumped by 13.5 m to an elevation of 1,900.5 masl, as shown in Figure 3.5 below. About 0.5 MCM of the embankment fill, or 10% of the total embankment volume, moved downstream. Figure 3.6 shows the failed downstream slope of the dam.

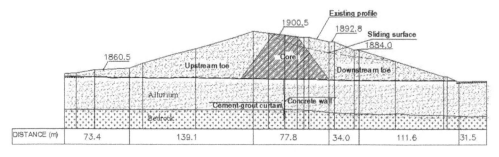

*Figure 3.5* Marmarik Dam after failure, prior to rehabilitation.

*Figure 3.6* Downstream slope of Marmarik dam just after failure, showing the failure scarps.

Immediately after the failure a local company was commissioned to investigate the dam. The investigation works were undertaken from 1975 to 1978 and concluded that the failure was caused by excess pore pressure in the clay core fill developed as a result of accelerated construction to meet the completion date; placement of the wet fill and the accelerated construction rate did not allow sufficient time for the high pore pressure in the core fill material to dissipate, resulting in settlements and slope failure.

After several unsuccessful rehabilitation attempts the dam was left in an unsafe condition for over 30 years. The river was diverted into a 3.2 m diameter diversion tunnel designed for a temporary condition, for a flood with a return period of 20 years. Since its completion, there have been three occasions during which the incoming flood exceeded the designed value, but the flood was absorbed in the reservoir storage volume without overtopping the dam. As there was a side weir at the outlet end of the diversion tunnel, which permanently maintained a minimum water level of 1.4–1.5 m in the lower section of the tunnel, the tunnel had never been inspected since its construction.

There are 142,000 people living downstream of the dam in 20 towns and villages at immediate risk from the dam failure.

b.  Safety Assessment and recommended works to improve safety

Safety assessment of the Marmarik dam was undertaken under DSP II and included:

- Safety inspection of the dam, the diversion tunnel and the landslides along the reservoir rim;
- Review of 1975–1978 investigation works;
- Topographic survey of the dam and the diversion tunnel;
- Supplementary ground investigations of the dam;
- Microseismic survey to establish site-specific seismic parameters;
- Landslides hazard assessment and landslide ground investigation of three landslides in the dam vicinity;

- Investigations of the diversion tunnel;
- Investigation of the efficiency of the foundation cut-off;
- Design of rehabilitation options
- Preparation of Dam Safety Plans

*Embankment*

Stability analysis of the embankment demonstrated that the dam, in its present condition, was stable as long as the reservoir level is kept 3 m below the crest. For an increased water level, the dam would be unsafe and had to be rehabilitated to improve its safety. The design challenge was to build a safe dam on top of the one that failed, using as much of the original dam as possible and to ensure that three landslides near the dam and reservoir do not impact the dam's safety. The following three options were developed for rehabilitation of the embankment:

- Option 1 – Reinstate the dam to the full height with the crest at 1,914 masl; Full Storage Level (FSL) at 1,911 masl, total storage volume 36 MCM.
- Option 2 – Reinstate the dam to evaluation of 1,905 masl; FSL at 1,902 masl, total storage volume 24 MCM.
- Option 3 – Reinstate the dam to evaluation of 1,889 masl; FSL at 1,886 masl, total storage volume 10 MCM.

Based on cost/benefit analyses, Option 2 has been adopted for implementation (see Figure 3.7).

Monitoring instruments that include piezometers, inclinometers, surface benchmarks and the seepage measuring weir have been recommended to be installed within the embankment and its foundations

*Landslide hazards*

Three potentially hazardous landslides were identified within the reservoir rim. There are as follows:

*Figure 3.7* Marmarik dam - upstream slope of the rehabilitated dam.

- N2 landslide some 250 m near the dam's right abutment
- N3 small landslide halfway up the reservoir, on the right-hand side
- N4 landslide, just at the upstream end of the reservoir, but largest in volume.

Inclinometers and surface benchmarks have been installed at the landslides N2 and N3. As for the landslide N4, it has been shown that if the landslide were to block the river, form a lake, and fail rapidly and completely, releasing about 2.33 million $m^3$ of water, it would cause the reservoir to rise about 1.9 m, leaving 1.1 m of freeboard. However, it is anticipated that should this event occur, the reservoir would be drawn down in anticipation of this event, which is covered in the Dam Safety Plans.

*Diversion tunnel, spillway, outlet works*

The following remedial measures were recommended to enable the operation of the diversion tunnel:

- Provision of a mass concrete plug upstream of the inlet pipes;
- Consolidation grouting downstream of the plug and back grouting of tunnel lining in areas of voids;
- Replacement of tunnel invert downstream of plug
- Drain holes to be incorporated in the invert to minimise hydrostatic loading

A new spillway is required at a lower level for the adopted Option 2 (see Figure 3.8). The outlet works required substantial refurbishment.

c.  Rehabilitation Works

These started in 2006 and were completed in 2012.

*Figure 3.8* Marmarik dam - new emergency spillway and bottom outlets.

### 3.3.6    Portfolio risk assessment

Portfolio Risk Assessment (PRA) was used under the DSPII to identify priority rehabilitation projects. The approach applied in the PRA was a semi-quantitative method
based on the probability and consequences of an event.[7] This method has been adopted
as the most appropriate method of assessing the safety of dams in Canada and the UK
where it is known as Failure Modes, Effects and Criticality Analysis.[8] The following
stages are required:

- identification of failure modes (instability, internal & external erosion etc)
- comparative assessment of the probability of failure
- comparative assessment of consequence or impact of failure

The result of the semi-quantitative method is the Risk Index, which is the product of
total impact and the risk score. The comparison of the Risk Indices provides useful a
ranking system to highlight the high-priority dams for remedial works.

Additionally, if the assessment is carried out where the assumed remedial works
have been repeated; then a reduction of the combined store will benefit the quantita-
tive assessment of the remedial works.

The Risk Indices, which illustrate the risk profile of Armenia's dams under DSP II
(see Figure 3.9) show that a reduction in risk is achieved by provision of the Emergency
Preparedness Plans and remedial works.

A reduction in Risk Index is a mean of assessing cost-effectiveness of the measures; the reduction is divided into the corresponding costs, both for structural (remedial works) and non-structural works (safety materials and EPPs).

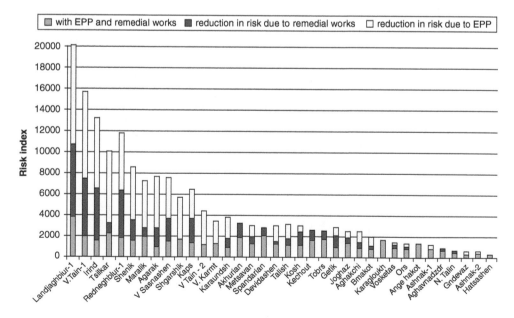

*Figure 3.9* Risk index for portfolio of dams.

## 3.3.7    Dam safety plans

Dam Safety Plans were produced under DSP II; the Dam Safety Plans comprised:

*   Instrumentation and Monitoring Plans,
*   Operation and Maintenance Plans
*   Emergency Preparedness Plans.

The Instrumentation and Monitoring Plans were provided for each dam. The instruments typically included:

*   staff gauge to monitor the level of the reservoir,
*   survey monuments installed at the crest and downstream berms to monitor settlements
*   piezometers, to monitor the phreatic surface and uplifts,
*   V-notch weirs, to measure seepages.

An automated alarm system was proposed to be installed for dams that presented a high hazard to the nearby communities; the alarm was expected to be triggered in the event of spillway discharge exceeding the design capacity.

The O&M Plans were also produced.

The EPPs were customised according to reservoirs' size and hazard potential. They comprised relevant technical information on dams, inundation maps produced based on dam break analysis, and potential safe havens. The EPPs were shared with the civil authorities and relayed to the emergency services.

In addition, dam safety equipment was also proposed to be provided and kept at regional depots or on sites of the major dams.

## 3.4    CONCLUSIONS AND CHALLENGES

Twenty dams out of 66 dams examined under the DSP II were classified as Extreme Risk Class; it was estimated that should the failure of these dams occur, over 500,000 people living downstream of the dams would be at risk.

The DSP II work demonstrated that by the implementation of both structural measures and EPPs, a significant safety improvement would be achieved for the dams. It was also demonstrated that implementation of EPPs alone, for the Extreme Risk Class dams, would play a significant contribution towards the safety of the population living downstream. The use of the semi-quantitative risk portfolio tool for undertaking a Portfolio Risk Assessment played an essential role in the prioritisation of implementation of remedial measures.

As a result of the two DSPs several deficiencies that made the dams unsafe were since eliminated; the most critical deficiencies that were identified were inadequate discharge capacities of spillways and embankment dams' slope instability, especially in seismic conditions. Both deficiencies were largely due to differences in requirements by the SNIPs, compared to the requirements of the more recent international standards.

Both DSPs also supported the preparation of Instrumentation and O&M plans, provision of the O&M equipment, improvements of access roads for safety interventions, preparation of EPPs and procurement and installation of telecommunication equipment, together with sirens. As a result, over 500,000 people (more than 15% of the entire population), living downstream of the rehabilitated dams, are now safe. 150,000 people were affected by the potential failure of Marmarik dam alone. Likewise, Armenia's infrastructure located downstream of the dams is no longer considered at risk.

### Challenges

There were some considerable challenges during the implementation of DSP II, many of which were related to communication, i.e. a language barrier, where the Armenian engineers worked in the Armenian and Russian languages only.

Also, the Armenian engineers have traditionally worked within the SNIPs rather than Western design and construction standards, which impacted the practicality and progress of the site investigation and laboratory testing and caused delays. This was also a challenge for the projects that were prepared for the implementation of construction works, which all had to get approvals from various local committees.

It also proved impossible to obtain reliable information for the majority of reservoirs in Armenia, due to the aftermath of the collapse of the former Soviet Union.

Also, difficult access to remote sites where the terrain is treacherous provided an extreme challenge.

Certain political events had a significant impact on the safety of some dams: the break up of the Soviet Union and the civil war that followed, put the unfinished Marmarik dam and Kaps dam at risk. Marmarik dam was rehabilitated in 2012, Kaps' rehabilitation is planned to start in 2022.

### Positives

On a positive side, as the country is a member of ICOLD, some Armenian colleagues involved in the DSPs, especially the ones working in the Project Implementation Unit, had a good knowledge of the WB dam safety requirements as well as ICOLD guides and bulletins, which made it easier to pass recommendations that largely differed from the accustomed SNIP requirements.

## NOTES

1  Armenia dam Safety project, Sawyer, J., Attewill, L. – BDS Conference 2004, Thomas Telford, London, 2004.
2  Marmarik Dam: Investigations and remedial works, Spasic-Gril, L., Sawyer, J. R. 2004: British Dam Society Conference, Canterbury.
3  Risks posed by an incomplete dam: Rehabilitation of Marmarik Dam in Armenia, Spasic-Gril, L., et al., ICOLD 2015 Stavanger, Norway.
4  NERC (1975) Flood Studies Report. London, UK: Natural Environment Research Council.
5  Schematic map of maiin tectonic features of the Caucasus and adjacent territories, Philip et al., 1989.
6  Seismic Design Standards of Republic of Armenia (SDSRA) II.2.02-94.
7  Portfolio Risk Management of Dams, Spasic-Gril, L., Santoro, D. – International Dam Safety Conference, 2019, Bhubaneswar, Odisha, India.
8  Failure Modes, Effects and Criticality Analysis (FMECA), https://en.wikipedia.org/wiki/Failure_mode,_effects,_and_criticality_analysis.

Chapter 4

# Case studies

## Eurasia/Western Asia – Georgia

### 4.1 ABOUT THE COUNTRY

Georgia is situated in Eurasia/Western Asia, mostly in the South Caucasus mountains, and some parts of the country are in the North Caucasus. The country is bordering Russia, Azerbaijan, Armenia, and Turkey (see Figure 4.1). The Likhi mountain range divides the country into two halves; the western area of Georgia is known as the Colchis while the eastern area is called Iberia. Georgia has a complex geographical setting, and the mountains isolate the north of Georgia (Svaneti) from the rest of the nation. The tallest mountain in Georgia is Mount Shkara, which is 5,068 m above sea level.

Modern Georgia became independent in 1991 during the *dissolution of the Soviet Union*.

**No. of people** – 3.7 million
**Area** – 69,700 km$^2$

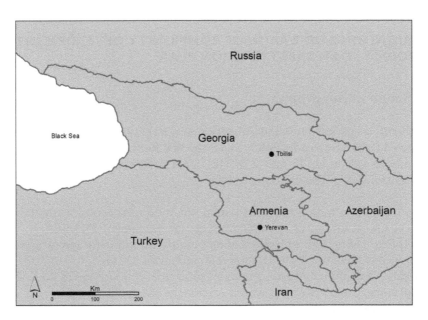

*Figure 4.1* Map of Georgia.

DOI: 10.1201/9780429320453-4

**Population Density** – 58/km$^2$
**Economy:**
GDP Nominal – (Total) $17.83 billion
GNI Nominal – (per capita) $4,740
**Climate** – Georgia's climate is very diverse, considering the size of the country; there are two different climatic zones in the west and east. The Caucasus mountains moderate the climate and protect the nation from the colder air coming from the north and also partially protect the region from the influence of dry and hot air masses from the south. Much of western Georgia lies within the northern periphery of the humid subtropical zone with annual precipitation ranging from 1,000 to 4,000 mm. The climate of the region varies with elevation, the lowland areas are relatively warm. The mountains and the foothills experience cool, wet summers and snowy winters. Eastern Georgia transitions from humid sub-tropical to continental, the climates in the regions are influenced by the dry air from the Caspian Sea from the east and the Black Sea from the west.

## 4.2   ABOUT THE DAMS

**Number of dams:** 50 (37 irrigation, 13 hydropower)
Most hydropower dams are large; the irrigation dams are both large and small in size. Most of the dams were designed and constructed in the 1950s and 1960s in compliance with the standards and practices accepted in the former Soviet Union.
**Laws and regulations**
There are no specific laws and regulations related to dams' safety. Standards used for dam design are largely based on the Soviet Design Norms (SNIPs).
**ICOLD membership**
Georgia is a member of ICOLD.

## 4.3   IRRIGATION AND DRAINAGE COMMUNITY DEVELOPMENT PROJECT - DAM SAFETY ACTION PLAN

### 4.3.1   Scope of the project

Four high-risk irrigation dams and a diversion weir structure on the Rioni River at Poti were selected to be assessed for safety under the WB financed Irrigation and Drainage Community Development Project (IDCDP), Dam Safety Action Plan (DSAP) undertaken in 2003–2004.[1,2] The dams are as follows:

- Sioni Dam, 84 m high embankment dam, the total reservoir storage of 325 MCM and the useful storage of 318.4 MCM, reservoir catchment 551 km$^2$
- Algeti Dam, 86 m high embankment dam, total reservoir storage of 65 MCM and the useful storage of 60 MCM, reservoir catchment 422 km$^2$
- Zonkari, 74 m high, embankment dam, total reservoir storage of 40.3 MCM and useful storage of 39 MCM· reservoir catchment 268 km$^2$
- Tbilisi dams, off-line storage
- River diversion scheme on the Rioni River at Poti.

*Figure 4.2* **Sioni dam on Ioari river.**

There is serious concern about capacity of spillways at Algeti and Sioni dams and seepages at Zonkari dam. Algeti dam is also located in the middle of a National Park, a protected area; its failure could impact the protected area and the towns downstream.

Tbilisi dam is located in the middle of the capital city, and its failure could pose extremely high risk to parts of Tbilisi.

The Rioni weir at Poti is an essential part of the Georgian infrastructure, and its collapse would cause a major disruption for the goods that arrive to the port of Poti (Figure 4.2).

The following work was undertaken under the DSAP:

1. Site inspections
2. Field Investigations – Topographic and bathymetric surveys, Ground investigations
3. Studies – Hydrology, flood routing, sedimentation, seismic hazard assessment, seismic analysis of dams, and stability analysis,
4. Rehabilitation works preliminary design and costing
5. Risk assessment
6. Dam safety plans (Operation & Maintenance, instrumentation, early warning systems, and emergency preparedness plan) (Figure 4.3)

### 4.3.2 Field investigations

For Sioni, Algeti, and Zonkari dams, the maps from the original design were used for the reservoirs, dam plans, and cross sections. Additional topographical survey was undertaken for:

* Tbilisi Reservoir – Area around the two concrete dams
* Algeti Dam – landslide area upstream of the dam at the right bank

*Figure 4.3* **Algeti dam on Algeti river.**

- Poti Diversion Weir – underwater survey to confirm the extent of the damage of the upstream and downstream apron of the stilling slabs

Supplementary ground investigation was undertaken at four dams and the Poti diversion weir.

In addition, investigation of the joints between the concrete facing slabs, at the left abutment at Sioni dam, at the locations where the seepage was noted, was undertaken.

## 4.3.3   Studies

### 4.3.3.1   HYDROLOGY

The dams, Sioni, Algeti, and Zonkari, on the rivers Iori, Algeti, and Patara Liakhvi, respectively, were designed and constructed during the Soviet period, in accordance with the SNIP, the Soviet standards applicable at that time for sizing of the capacities of the outlet structures. Hydrological studies were undertaken under this project to estimate the adequacy of the spillways when the design floods are selected in accordance with current international practice. The hydrological studies were carried out in the following steps:

- Selection of the return period of the design flood according to the hazard posed by a dam failure
- Collection of climatic and hydrological data
- Determination of the flood hydrograph for the selected return period for each dam.

*Table 4.1* Selection of design flood according to SNIP

| Dam class | Design flood | Check flood |
|---|---|---|
| 1 | 1 in 1,000 year | 1 in 10,000 year |
| 2 | 1 in 100 year | 1 in 1,000 year |
| 3 | 1 in 33 year | 1 in 200 year |
| 4 | 1 in 20 year | 1 in 100 year |

i.  Selection of the return period of the design flood

**SNIP**
The selection of the design flood is based on the dam class, which, in turn, is based on a dam height and reservoir volume only. The dam classes are as follows (Table 4.1):

Based on the above SNIP classification, the following are the design floods:
- Algeti Dam – Class 2, design flood 1 in 100 years, check flood 1 in 1,000 years
- Sioni Dam – Class 2, design flood 1 in 100 years, check flood 1 in 1,000 years
- Zonkari Dam – Class 2, design flood 1 in 100 years, check flood 1 in 1,000 years

**ICOLD**
Based on the ICOLD classification (see Section 2.3.2), the following are the design floods:
- Algeti Dam – Class IV, Extreme Risk, design flood 1 in 10,000 years, check flood PMF
- Sioni Dam – Class IV, Extreme Risk, design flood 1 in 10,000 years, check flood PMF
- Zonkari Dam – Class IV, Extreme Risk, design flood 1 in 10,000 years, check flood PMF

**Recommended design floods**
Although there is a considerable difference in the choice of the return period between SNIP and ICOLD, the design floods based on the ICOLD were recommended bearing in mind the large number of people living downstream of the dams and the extreme risk the dam break represents.

ii.  Determination of the design flood
This was done using three methods, as follows:

- Georgian method", based on SNIP techniques, which is, in general, used in Georgia.
- Regional method, described in Section 3.3.1
- Estimate of the Probable Maximum Flood (PMF)

iii.  Summary and recommendation
The floods recalculated for Sioni and Algeti dams using the Georgian method show good agreement with the regional method, for both 1in1,000 years and 1 in 10,000 years floods. The floods for the Zonkari dam estimated by the regional method are lower than the ones estimated by the Georgian empirical method developed for the small catchments in the Caucasus.

The original design was carried out for the inflow floods with 1 in 100 years return period and checked for the 1 in 1,000 years floods. Such low flood return periods for the high-risk dams are not acceptable, and it was therefore recommended that the hydraulic studies are carried out for 1:10,000 floods and checked for PMF.

### 4.3.3.2   Sedimentation

Although it was not anticipated during the design that any of the reservoirs would be filled by sedimentation within the first 100 years of their life, an independent sediment check was undertaken under the Dam Safety Assessment. Based on these estimates, it can be concluded that the sedimentation is likely not to pose any risk to Algeti and Sioni reservoirs during their lifetime. However, that cannot be concluded for the Zonkari reservoir which might have a useful life of less than 100 years. It is recommended that a bathymetric survey is carried out at Zonkari reservoir, and estimates of sediments yield are reviewed based on the survey data. Based on the results, some long-term sedimentation mitigation measures might be necessary.

### 4.3.3.3   Seismic studies

Seismicity of the Caucasus region is described in Section 3.3.2.

Due to an extremely high seismic activity in the region, a special consideration was given when selecting seismic design parameters for checking seismic performance of the four dams and the Rioni weir.

It was previously recognised by others that the SNIP seismic maps and the seismic coefficients proposed for Georgia, used in the original design, were not in line with the more recent seismic requirements. A new Georgian Seismic Code was established in 1999 based on Probabilistic Seismic Hazard Assessment (PSHA). A set of maps for macroseismic intensity and peak ground acceleration (PGA) for 50 years exposure time and 0.5%, 1%, 2%, 5%, and 10% probability of exceedance were constructed for the Type II foundation soils (weathered rock and dense alluvium).[3]

The PSHA maps provide the following PGAs for the four dams and the Rioni weir, the Type II soils.

*Table 4.2* **PGA for different return periods/probability of exceedance**

| Site | Probability of exceedance in 50 years | | | | |
| | 0.5% | 1% | 2% | 5% | 10% |
| | Return period | | | | |
| | 1:10,000 | 1:5,000 | 1:2,500 | 1:1,000 | 1:500 |
| --- | --- | --- | --- | --- | --- |
| Tbilisi | 0.25 | 0.170 | 0.137 | 0.104 | 0.07 |
| Sioni | 0.352 | 0.307 | 0.248 | 0.179 | 0.15 |
| Poti | 0.22 | 0.19 | 0.15 | 0.11 | 0.06 |
| Zonkari | 0.4 | 0.323 | 0.263 | 0.183 | 0.15 |
| Algeti | 0.25 | 0.198 | 0.16 | 0.11 | 0.10 |

The PGA values for 1:500 and 1:10,000 return period have been recommended to be adopted for the OBE and the MDE, respectively, when checking the seismic performance of the dams.

For all the sites but Poti, the foundation materials have been classified as Type II soils. Therefore, the PGAs shown in Table 4.2 above have been applied for seismic assessments. However, for the Poti weir foundation materials are loose, saturated silts and sand, and therefore for this site, a site-specific hazard assessment was carried out. The results showed a magnified PGA with the OBE = 0.4 g and the MDE = 0.6 g. These values were used when the stability of the Poti weir was checked.

### 4.3.4  Typical deficiencies and rehabilitation works proposed for four dams

The main defects for the four dams have been grouped as follows:

1. Defects of the Dam – lack of sufficient freeboard (Algeti and Sioni dams), factor of safety in seismic conditions (<1.0 for Sioni dam, but deformations are negligible)
2. Defects of the Spillway (Including Hydromechanical Equipment) – insufficient spillway capacity for Algeti, Sioni and Zonkari dams
3. Defects of the Outlet Works (Including Hydromechanical Equipment) – all four dams

It was recommended not only to undertake remedial works as soon as possible but also to keep the reservoir levels for the Algeti, Sioni and Zonkari dams at reduced levels, until the remedial works are completed.

### 4.3.5  Safety aspects of the Rioni diversion weir

#### *4.3.5.1  Description of the diversion structure*

The diversion structure across the Rioni river was constructed to protect the town of Poti from frequent flooding. Poti is the third largest town in Georgia and is currently the busiest Georgian port on the Black Sea. The structure is located about 2 km from the Black Sea. The design was done between 1948 and 1951 in accordance with SNIP and the structure was constructed from 1952 to 1959.

The diversion comprises two independent weir structures (see Figure 4.4), namely:

- the left bank regulator which discharges 400 m³/s into a canal that goes through the town of Poti and
- the main weir, which contains ten openings that discharge up to 400 m³/s each and divert the river flood flow towards the Black Sea, away from the town of Poti.

Since its construction, the main weir has also been used as the only road bridge across the Rioni river that leads to Poti. During the 1990s, the traffic across the bridge intensified and the bridge, also known as the "Euro – Asian bridge", has been used for transport of heavy goods to and from the Port of Poti.

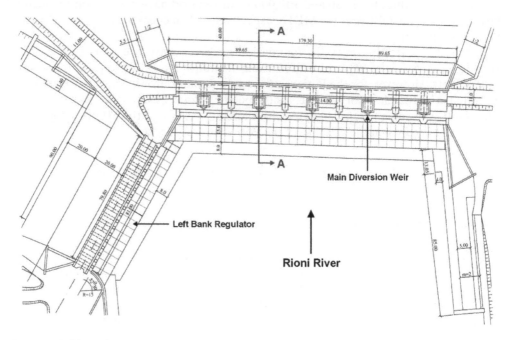

*Figure 4.4* **Plan of diversion structures.**

### 4.3.5.2   History of scour problems

Scour problems related to the main diversion structure have been reported for the last 50 years. It was indicated that scouring of the downstream pool started immediately after the facility was put into operation.

Between 1982 and 1983 some remediation measures were undertaken, which also included installation of a sheet pile wall; however, the sheet pile wall installed 80 m downstream from the structure increased the scour depth by 25%–28% and encouraged the scour line to move closer to the main structural foundations.

Observations made in 1998–1999 revealed that the scour depth was 6–10 m and extended to the edge of the stilling slabs.

Another attempt to rehabilitate the structure at the beginning of 2001 led for the design to be put on hold several times due to a lack of funds.

An underwater survey was undertaken in 2003; it showed that the erosion had progressed close to the main structural foundation of the central piers, putting the structure at a verge of breakdown, endangering the safety of the population in the town of Poti and nearby villages, as well as the important transportation links and sustainability of the Black Sea coastal zone.

Emergency rehabilitation works were undertaken in 2004, which included surveys, investigations, studies, and implementation of the emergency remedial works.[4] These works are described in the sections below.

### 4.3.5.3    Inspections and surveys under DSAP

The following inspections and survey were undertaken in 2004:

• Inspection of the hydromechanical and electrical equipment of the main diversion structure and the left bank regulator
• Inspection of records of geodetic survey
• Underwater survey by divers and the bathymetric survey
• Supplementary geotechnical investigation

Due to aged hydromechanical and electrical equipment and a lack of maintenance funds, many parts of the equipment were out of operation, i.e. the main regulator could only discharge river flows through the central gates.

The underwater survey by divers and the bathymetric survey identified that the scour extended to the edge of the main structural foundation in the area of the central sections and pier No 5 where 3–4 m deep voids were encountered (see Figure 4.5).

The following was found:

• The stilling slabs had been completely demolished in the central sections and partly in the remaining sections.
• The downstream apron had been completely damaged and largely disappeared
• The upstream apron had been damaged in the central part; the last 5 m of the apron slab had been damaged over a length of 100 m

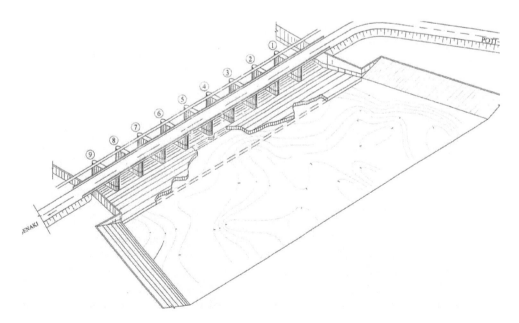

Figure 4.5 **Scour of the weir structure.**

### 4.3.5.4    Studies under DSAP

Upon completion of surveys, studies were undertaken, which comprised hydraulic and seepage and structural analyses. To understand the mechanisms which caused the failure of the downstream apron and the stilling slabs, the original design was studied and compared with good hydraulic engineering practice. The impact of the operational regime of the weir was also studied as well as the rehabilitation measures that were implemented in the early 1980s.

The causes of failure of the downstream apron and the stilling slabs could be grouped under three main headings:

*   inadequate original design
*   inappropriate operation of the gates
*   inappropriate design of the rehabilitation works implemented in 1982–1983- the sheet pile wall installed 80 m downstream from the structure increased the scour depth by 25%–28% and encouraged the scour line to move closer to the main structural foundations.

### 4.3.5.5    Emergency remedial works

Following the conclusion of the analyses, emergency rehabilitation works were recommended under the DSP. The emergency rehabilitation works were carried out under water during a low flow season, with a complete closure of all ten gates. The flow of less than 400 m³/s had to be diverted via the left bank regulator into the Poti canal. The following were the recommended actions:

*   Prohibit the use of the diversion structure by heavy traffic until emergency rehabilitation works are implemented. This was recommended by the DSAP and was endorsed and implemented by the Georgian Government in March 2004.
*   Rehabilitate hydromechanical and electrical equipment on the main regulating structure and the left bank regulator
*   Clear the Poti canal so that flows of up to 400 m³/s could be discharged through the left bank regulator
*   Backfill the voids close to the structural foundation, in the areas of the central sections and the pier No 5, that are associated with erosive processes and are likely to have an impact on the structural stability. This was carried out by placing sand-cement-clay mix underwater through tremies installed through voids identified in the broken stilling slabs, close to the structure. Prior to the backfilling a site trial was carried out to check the exact mix and the pumping pressures. Quality of backfilling works was checked by drilling and sampling of the backfilling.
*   Reinstatement of the stilling slabs and the downstream apron in the areas where the slabs and the apron are damaged. This was carried out using Maccaferri gabion baskets (2 × 2 × 1 m) which were assembled on the river bank, dropped into position from barges and subsequently interconnected underwater by divers. A geofabric was placed along the outer sides of the gabion baskets to prevent upward migration of the fines from the foundation soils into the gabions. Prior to the

placement of the gabions, a bedding layer of sand and gravel was placed over the scoured foundation to level the area. In places where the gabions were to be placed over broken slabs close to the main structure, smaller and more flexible cylindrical gabions were used. The cylindrical gabions were 1 m long and 0.65 m in diameter. They were also used at the ends of the "gabion structure" to provide a transition between the gabions and the natural river bed. Divers were employed to ensure that the gabions were properly placed and interconnected. An independent team of divers was employed by the implementing agency, Project Coordination Centre of the WB, to control the quality of the underwater works.

The implementation of emergency rehabilitation works started in September 2004 with the rehabilitation of the hydromechanical and electrical equipment on the main regulating structure and the left bank regulator and clearing of the Poti canal. These works were delayed and had impact on the programme of placement of the gabions. Between February and March 2005 the works were undertaken with full closure of the gates, except for a short period of 3 days at the end of February when the flood flow was about 1,400 m$^3$/s and gates 1, 8, 9 and 10 were opened. After passing of the 1,400 m$^3$/s flow at the end of February, a bathymetric survey was carried out which demonstrated that flow did not pose any damage to the works installed. The emergency rehabilitation was completed in February 2006.

### 4.3.6   Risk assessment

The approach used in the assessment of the risk level for the four dams and the Poti weir is a semi-quantitative method described in Section 3.3.6, where a risk profile for a portfolio was measured by Risk Indices estimated for:

- the existing condition,
- after implementation of remedial measures only,
- after implementation of EPP only, and
- after implementation of both the remedial measures and the EPP.

**Existing Risk Profile**
The highest risk was obtained for the Poti weir. The risks for Algeti, Sioni and Zonkari are almost identical. Comparatively with the other four structures, the risk at Tbilisi is the lowest.
**Reduction in risk index due to remedial measures**
The biggest risk reduction due to the implementation of remedial measures was obtained for the Poti weir. A significant risk reduction was also obtained for Algeti and Sioni after the construction of the wave wall (Algeti) and widening of the spillway (Sioni).
**Reduction in risk index due to implementation of EPPs**
Implementation of EPPs can significantly reduce the existing risk score if they are fully effective and maintained.
**Reduction in risk index due to combined remedial works and EPP**
The risk analyses showed that the reduction in risk due to implementation of the structural measures and the EPPs is the biggest for Poti weir. The risk reduction for

Algeti, Sioni and Zonkari dams are almost identical, while the risk reduction for Tbilisi is the lowest.

The analyses also showed that implementation of effective EPPs could significantly reduce the risk for all five structures, but the biggest risk reduction will be for Poti and Tbilisi.

**Conclusions and recommendations**

The following recommendations were based on the risk analyses:

- Immediately undertake Priority I works at the Poti weir. As soon as it is practically possible develop an EPP and install an effective EWS;
- Algeti, Sioni, and Zonkari dams have a similar risk level and the same priority of the implementation of the remedial works. However, remedial works should be first implemented at Algeti and Sioni as the risk of overtopping for these two dams is high due to insufficient freeboard allowance. It is also recommended to develop an EPPs and install an EWS at these three dams;
- It is recommended that the Tbilisi reservoir FSL is kept at the present level; It is recommended to develop an EPPs and install an EWS at this dam.

### 4.3.7   Dam safety plans

There are currently no monitoring instruments on any of the dams; they are either stolen or are non-operational, no monitoring records exist except the records of daily reservoir levels (and seepage measurements for Zonkari dam only).

There are no O&M manuals and no EPPs. The primary responsibility of the O&M staff is to open and close gates and valves as necessary to release water at the rates needed to support irrigation. Local O&M staff are involved in routine maintenance, although the annual budget allocated to each dam is negligible.

There are no regular inspections by qualified engineers; the dams are only inspected by the engineers from the Design Institute from Tbilisi when the local staff raise a concern about some matter.

This situation is not acceptable bearing in mind the hazard the dams pose to the downstream population. It is therefore recommended that for each dam the following is prepared which will be a part of the Dam Safety Plans:

- Instrumentation Plan
- O&M Plan, and
- EPP

### 4.4   CONCLUSIONS AND CHALLENGES

All dams and the Poti weir, investigated under the DSAP project were classified as an Extreme Risk Class dams; it was estimated that, should failure of these dams occur, over thousands of people living downstream of the dams would be at risk.

The DSAP work demonstrated that by the implementation of both structural measures and EPPs, a significant safety improvement would be achieved for the dams and the Poti weir structure. It was also demonstrated that implementation of EPPs alone, would play a significant contribution towards the safety of the population living downstream of the Tbilisi dam and the Poti weir.

**As the dams were designed in accordance with the SNIP norms, which largely underestimated design and safety check floods,** the most critical deficiencies that were identified were inadequate discharge capacities of spillways and inadequate freeboard.

No monitoring instruments exist on dams, which is unacceptable, bearing in mind the Risk Class of the dams. Also, negligible sums have been allocated by the Government for dams' maintenance, which too, is unacceptable. The DSAP project supported the preparation of O&M plans, provision of the O&M equipment, revision of EPP) and procurement and installation of telecommunication equipment, together with sirens. With these measures in place, the risk to the downstream population due to a potential failure of dams and the Poti weir would be significantly reduced.

**Challenges**

As in Armenia, the engineers in Georgia have traditionally worked within the SNIP rather than Western norms, which impacted the practicality and progress of the site investigation and laboratory testing. However, the local experts that worked within the Project Implementation Units, had a good knowledge of the WB dam safety requirements as well as ICOLD guides and bulletins which made it easier to pass recommendations that largely differed from the accustomed SNIP norms.

## NOTES

1  Spasic-Gril, L., Dam safety project in Georgia, Dams and Reservoirs, July 2005, 15(2).
2  Spasic-Gril, L., Dam safety project in Georgia, Georgia, SANCOLD Conference, Johannesburg, 2011.
3  Seismic hazard assessment of Georgia (probabilistic approach) Chelidze, T., et al., 1999.
4  Spasic-Gril. L., Emergency underwater rehabilitation of the Poti main diversion weir, Georgia,14th BDS Conference, Durham, September 2006.

# Chapter 5

# Case study

## Central Asia – Tajikistan

## 5.1 ABOUT THE COUNTRY

The country, shown in Figure 5.1, is situated between Kyrgyzstan and Uzbekistan to the north and west, China to the east and Afghanistan to the south. Mountains cover most of Tajikistan's area, the most prominent are the Pamirs and the Alays. Central Asia's other mountain range Tian Shan is on the border with northern Tajikistan. The Pyanj river in the south, at the border with Afghanistan, and the Fergana valley in the north are areas of agricultural and industrial activity.

Tajikistan became an independent sovereign nation in 1991 when the Soviet Union *disintegrated*.

No. of people – 9.5 million
**Area** –143,100 km$^2$
**Population Density** – 66/km$^2$
**Economy:**
GDP Nominal – (Total) $7.35 billion
GNI Nominal – (per capita) $1,030
**Climate** - The climate is sub-tropical and continental, although there are semi-arid areas with desert. The climate changes with the elevation, the Fergana Valley is shielded from the Arctic air but temperatures remain freezing for more than 100 days of the year. In the lowlands in the south-west, the climate is sub-tropical and it is ideal for farming, and their average temperatures range from 23°C to 30°C in the summer, and −1°C to −3°C in the winter. However, in the eastern Pamirs, the summer temperature is only 5°C–10°C and can drop to as low as −15°C to −20°C in the winter months.

Most of the precipitation is in the winter months and the rivers are largely controlled by snow and glacier melt.

## 5.2 ABOUT THE DAMS

**No. of Dams**: there are nine man-made dams in Tajikistan; Six dams, all mainly for hydropower, are in a cascade on the Vakhsh River, which is a tributary of the Pyanj River, and forms the upper part of the Amu Darya River, that flows into the Aral Sea. A list of the dams on the Vakhsh Cascade is given in Table 5.1 below.

DOI: 10.1201/9780429320453-5

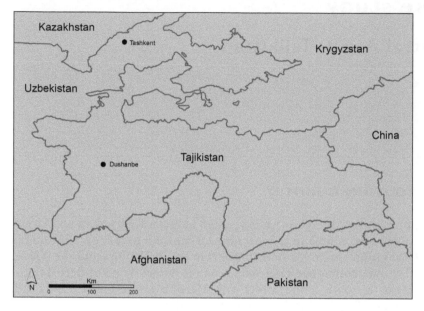

*Figure 5.1* Map of Tajikistan.

*Table 5.1* Hydropower dams on Vakhsh cascade

| Dam | Current installed capacity (MW) | Commissioning | Major rehabilitation or upgrade |
|---|---|---|---|
| Rogun | 3,600[a] | 2028 | Under construction |
| Nurek | 3,000 | 1972–1979 | Ongoing |
| Baipaza | 600 | 1985–1986 | Feasibility study for rehabilitation works started |
| Sangtuda-1 | 670 | 2008–2009 | Not required |
| Sangtuda-2 | 220 | 2011–2014 | Not required |
| Sarband[b] | 240 | 1962–1963 | Ongoing |

[a] 3,600 MW is a full capacity, once the dam is fully constructed; currently the capacity for early generation during construction is 400 MW with two temporary units and lower head.
[b] Dam also known as Golovnaya.

The top two dams on the Vakhsh cascade include the 300 m high Nurek dam, the second tallest dam in the world, and 335 m high Rogun dam, currently under construction, to be the tallest dam in the world.

In the 1950s and 1960s, there was a plan to construct the 335 m high Rogun Hydropower Dam, with an underground powerhouse with a capacity of 3,600 MW, located some 70 km upstream on the Nurek Dam. The two dams were envisaged to work in a cascade. The construction of Rogun started in 1976 but stopped in 1991, due to the break up of the Soviet Union and the civil war that followed. In 1993, a large flood destroyed a cofferdam, which, at the time, had reached a height of 60 m. The construction works at Rogun restarted in 2017 (see Figure 5.2).

*Figure 5.2* **Rogun dam currently under construction.**

A portfolio of dams also includes a 500–700 m high Usoi natural dam (see Section 5.4 below), the highest natural dam in the world. Usoi dam retains 60 km long Lake Sarez, with a total volume of reservoir of 17 km$^3$.

**Law and regulations**

There are no specific laws and regulations related to dams' safety. Standards used for dam design are largely based on the Soviet Design Norms.

**ICOLD membership**

Tajikistan has been a member of ICOLD

## 5.3   NUREK DAM

### 5.3.1   Background

Nurek HPP, in Figure 5.3 below, located on the Vakhsh River, 80 km southeast of the capital Dushanbe, is the backbone of the Tajik power system, as it generates over 70% of the electricity consumed in the country.

Construction of the dam started in 1961, while Tajikistan was a part of the Soviet Union, and when completed in the 1970s, it became the tallest dam in the world, 300 m high. It is currently the second tallest man-made dam in the world, after being surpassed in 2013 by Jinping-I Dam, which is 305 m high.

The Nurek HPP has nine Francis turbines; the initial installed capacity of 2,700 MW (300 MW per unit) was increased in 1988 to 3,000 MW through an upgrade of the turbines.

The Nurek reservoir was also designed to provide water for *irrigation* of 70,000 ha of *agricultural* land.

An estimated 5,000 people were resettled from the dam's flooding area prior to construction.

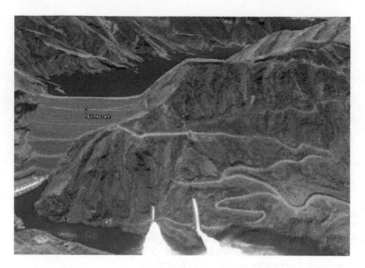

*Figure 5.3* Nurek dam with two tunnel spillways operating.

It is estimated that over 50,000 people living downstream of the dam could be at risk, if the dam fails.

## 5.3.2   Main project features

The principal project data are presented in Table 5.2.

**Embankment dam and "seismic belts"**

The dam is an embankment with rockfill shoulders and a central clay core (see Figure 5.4). There are two filters on the upstream and downstream sides of the clay core. The clay core is founded on a concrete slab.

Due to a high seismicity of the area and a unique heigh of the dam, one of the major concerns during the design was a seismic performance of the dam during earthquakes, especially the magnification of the seismic bedrock acceleration towards the dam crest. Even a physical model of the embankment was performed, and a decision was made to introduce "seismic belts" in the upper part of the crest to enhance the dam's seismic stability.

The seismic belts were constructed within the upstream (u/s) and downstream (d/s) shoulder of the embankment. They were placed as follows:

- On u/s Slope at Elevations – 912 to 915[i], 894, 876 and 855 masl;
- On d/s slope at elevation 912–915 masl.

According to the original design report, the seismic belts comprise discrete concrete blocks (Figure 5.6), interconnected by reinforcement, placed within the rockfill material to provide confinement to the rockfill and limit seismic deformations. The blocks

---

i   In the gallery at the top of the clay core the reinforcement that connects the u/s and d/s seismic belts is placed at the top of the core, i.e. at 915 masl. The blocks placed within the u/s and d/s shoulders then slope towards the elevation 912 masl.

*Table 5.2* Nurek dam- main project data

| Reservoir | |
|---|---|
| Catchment area | 30,700 km$^2$ |
| Reservoir volume at FSL | 10,500 Mm$^3$ |
| Reservoir length at FSL | 70 km |
| Dead storage at LDL | 6,000 Mm$^3$ |
| Full storage level (FSL) | 910 masl |
| Lowest drawdown level (LDL) | 857 masl |

| Dam type | Rockfill embankment with clay core |
|---|---|

| | |
|---|---|
| Dam crest level | 920 masl |
| Dam crest length | 714 m |
| Dam crest width | 20 m |
| Freeboard | 10 m |
| Height of embankment | 300 m |
| Core crest level | 915 masl |
| Upstream slope | 1:2.25 |
| Downstream slope | 1:2.2 |

| Flood discharge capacity | |
|---|---|
| Design flood, 1 in 1,000 year (0.1%) | 4,040 m$^3$/s |
| Check flood, 1 in 10,000 year (0.01%) | 5,400 m$^3$/s |
| Spillway discharge capacity, (two tunnel spillways) | 4,040 m$^3$/s |
| Max. discharge through turbines | 1,350 m$^3$/s |

| Hydropower equipment | Units | Quantity | Type |
|---|---|---|---|
| Hydraulic turbines | No | 9 | PO-310/957-B-475 |
| Capacity[a] | MW | 341 | |
| Diameter | m | 4.75 | |
| Design head | m | 230 | |
| Design head water discharge | m$^3$/s | 158 | |
| Switchyard | No | 1 | 220 kV |
| Switchyard | No | 1 | 500 kV |

[a] Capacity has been increased after the 1988 upgrade.

were placed at 10 m centres and are not connected in the longitudinal direction. They are also not anchored to abutments.

The u/s and d/s blocks at elevation 915 masl have been connected by reinforcing bars that are visible in the top gallery (Gallery III), see Figures 5.5 and 5.6, below.

Some authors[1] disagree about the necessity of seismic belts. Nevertheless, the dam has performed well since its construction with no significant displacements measured during earthquakes.

**Spillways**

There are two spillway tunnels on the Left Bank of the dam (Figure 5.7) designed for an evacuation capacity of 4,040 m$^3$/s (i.e. the Design Flood with 0.1% annual probability), as per the SNIP norms. The additional evacuation capacity of 1,350 m$^3$/s (the remaining of the Safety Check flood, with 0.01% annual probability) is to be provided by discharging through nine turbines.

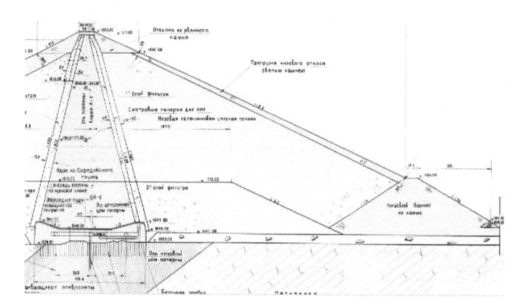

*Figure 5.4* **Cross section through the embankment.**

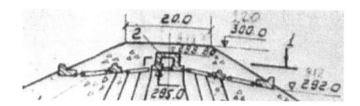

*Figure 5.5* **Gallery III and the upper seismic belt (original design drawing).**

### Seismicity of the site

Nurek dam is located within an exceptionally complex tectonic area, in the region of the Indian and Eurasian tectonic plates collision, and more accurately, just north of the western extent of the Himalayan range. The dam is in the Tajik Depression, in the Vakhsh range, which is a portion of the active deformation zone resulting from the Cenozoic collision between the Indian and Eurasian tectonic plates. Crustal shortening between the Pamir and Tien Shan mountain ranges is an important consequence of the convergence. The Pamir and Tien Shan mountain ranges are marked by significant seismicity, involving crustal earthquakes concentrated along fault systems that form their boundaries.

The original Nurek Dam design was undertaken in accordance with the SNIP norms, to the Medvedev-Sponheur-Karnik Seismic Intensity Scale, MSK – 64 Intensity scale IX. (The dam falls into Intensity zone VIII, but the intensity was increased to IX due to the importance of the structure). The peak ground acceleration associated with MSK-64 Intensity scale IX varies from 0.35–0.6 g. However, this acceleration is typically related to a return period of 475 years.

*Figure 5.6* **Concrete blocks on d/s slope of the dam, part of the seismic belt at elevation 912 masl.**

### Seismic Monitoring at Nurek Dam

In the 1980s, the seismic monitoring instrumentation at Nurek comprised 12 monitoring instruments, placed in the dam and the abutments; these were three-component Strong Motion Accelerometers (SMA). Seismic performance of the dam was assessed to be very good; however, most of these instruments are not now functioning.

### Reservoir triggered seismicity

Filling of the Nurek reservoir has been accompanied by a significant level of the Reservoir Triggered Seismicity (RTS).[2] Due to the complex geological and tectonic setting in the area, with high seismic activity, seismological and geological studies and monitoring of the reservoir started in 1955, well before the reservoir filling. This made it possible to determine the extent of changes and the rate of seismic activity due to the reservoir filling.

More than 1,800 earthquakes ($1.4 > M > 4.6$) occurred during the first 9 years of filling. This was more than four times the average rate of activity in the region before the start of the filling. However, the monitored earthquakes had a maximum magnitude of $M = 4.6$. The RTS was monitored some 10 years after the full impoundment when the reservoir triggered seismic activity ceased.

### Sedimentation

A study was done in 2010 to look at Nurek reservoir sedimentation.[3] The study concluded that over 35 years of Nurek reservoir operation practically the entire solid

*Figure 5.7* Two spillway tunnels on the Left Bank.

runoff of the Vakhsh River was accumulated in the reservoir, indicating a loss of storage of about 1% per year. The estimated loss of storage at the time was about 35%, mainly in the upper parts of the reservoir.

**Monitoring Instrumentation**

A large number of instruments have originally been installed in the dam and appurtenant structures for monitoring. These instruments recorded a good performance of the dam in terms of:

- settlements,
- seepage
- seismic displacements, etc.

However, a good number of instruments are now out of order and need to be replaced.

### 5.3.3   Dam safety and reservoir management project and Nurek HPP rehabilitation project

*5.3.3.1   Dam safety and reservoir management project*

The safety of Nurek Dam was addressed in early 2000 under the Aral Sea Basin Program, Water and Environmental Management Project (WAEMP), Component C – Dam Safety and Reservoir Management.

The WAEMP, supported by a variety of donors, was aimed at addressing the root causes of overuse and degradation of the international waters of the Aral Sea Basin comprises the Syr Darya and Amu Darya River basins. The Dam Safety and Reservoir Management component covered the safety of ten largest dams in the Aral Sea Basin, including Nurek Dam. For all ten dams the project recommended the implementation of rehabilitation measures, installation of adequate instrumentation and early warning system, preparation of emergency preparedness plan and training. These recommendations have also been applied to Nurek Dam. The project also included Lake Sarez (see Section 5.4).

## 5.3.3.2   Nurek HPP rehabilitation project

No capital rehabilitation of Nurek HPP was undertaken since the start of its operation in the late 1970s. After more than four decades of operation, the dam and hydropower plant required major rehabilitation of its generating units and a series of works to ensure its continued safe operation. In recent years, the available capacity of the Nurek HPP is some 2,300 MW as Unit 8 has been out of operation since 2011; increasing the Nurek HPP generating capacity is the key to achieving energy security in the country.

Feasibility study for Nurek HPP Rehabilitation was commissioned in 2014 with the financial assistance of the WB. In parallel, also under the WB funding, a Sedimentation Study was undertaken. Both, the feasibility study and the sedimentation study, were completed in February 2016, with some additional works also done in 2017.

The rehabilitation of the Nurek HPP will be carried out in two phases. The first phase is expected to comprise the rehabilitation of three generating units and critical dam safety works. The remaining six units will be rehabilitated in the second phase of the project. The first phase of the rehabilitation project has already been commissioned, and it is ongoing.

The critical dam safety works, supported by the project could be summarised as below:

**Safety of dam under PMF**

It has been identified under Dam Safety Studies that the PMF discharge, which is the most appropriate to use as the Safety Check Flood, could be largely more than 0.01% flood adopted in the original design. It has also been recognised that Rogun Dam, currently under construction 70 km upstream of Nurek, has been designed to store the PMF (based on recent studies for Rogun dam) so that the downstream releases are always less than the original Safety Check flood of 0.01%; therefore, the long term safety of Nurek against the PMF is assessed to be adequate, once Rogun dam is fully constructed. Remediation measures are envisaged to be put in place, that would ensure Nurek dam is safe against overtopping during a PMF in the period when Rogun dam is not fully constructed. These measures are currently been discussed under the Nurek HPP Rehabilitation Project.

**Seismic performance**

As indicated above, the design accelerations originally selected for the Nurek dam were based on the SNIP norms and are related to a return period of 475 years. This is not adequate, based on the International Guides (ICOLD etc); therefore, under the Nurek HPP rehabilitation project, a PSHA has been undertaken to determine PGAs for return periods of 145 years (OBE), 475 years and 10,000 years (SEE). Seismic performance of Nurek Dam under SEE and potential rehabilitation of the seismic belts are currently been discussed under the Nurek HPP Rehabilitation Project.

**Sedimentation**

Sedimentation study was undertaken in 2015–2016[4] under the Nurek HPP Rehabilitation Project. The previously estimated rate of sedimentation, which indicated a loss of storage of about 1% per year, was confirmed.

Since there is no significant lateral inflow between Rogun and Nurek reservoirs, the same average annual sediment deposition rate may also be assumed for the Rogun reservoir. Thus, after the commissioning of the Rogun dam, sedimentation at Nurek will significantly reduce, thereby increasing the remaining economic lifetime of the Nurek dam.

**Investigation of Stability of the Left Bank and Rehabilitation of spillway tunnels**

Both spillway tunnels are located at the Left Bank, which has a complex geology, with a dome of salt and gypsum being uplifted along active faults. Figure 5.8 shows a schematic geological model of the Left Bank, with both tunnels intersected by an active fault.

A slow settlement of few mm/year has been recorded along the fault, which provoked the rupture in the tunnels. Geotechnical Investigation campaign has currently been undertaken at the Left Bank. Monitoring instruments have also been installed in boreholes at the Left Bank. Information from both, the Geotechnical Investigation and Monitoring, will provide data for stability assessment of the Left Bank stability, which will be undertaken. Also, rehabilitation of the tunnels is currently under discussion.

**Monitoring Instrumentation**

Refurbishing and upgrading monitoring instruments and management system to improve the collection and analysis of the safety monitoring data is also considered under the Nurek HPP Rehabilitation Project.

**EPP**

It is proposed to update the Emergency Preparedness Plan, Operation and Maintenance Plan, and the Instrumentation Plan for the dam.

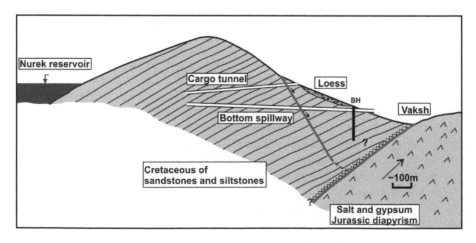

*Figure 5.8* Schematic geological model of the Left Bank.

**Training and Capacity Building**

This is currently been undertaken under the Nurek HPP Rehabilitation project. The training also included training on transboundary issues, examples of transboundary rivers and dams and mitigation measures typically required.

## 5.4 USOY DAM AND LAKE SAREZ

### 5.4.1 Introduction

Usoy dam and Lake Sarez are situated in the extremely remote and rugged terrain of the Pamir mountains in Tajikistan (Figure 5.9 below). They were formed in February 1911 when a magnitude 7.4 earthquake triggered a massive landslide, which dammed the Murgab River, in the Pamir Mountains, burying the village of Usoy.[5,6] The natural dam is between 550 and 720 m high with a minimum crest elevation of 3,320 masl and the water level of the lake thus formed has been rising ever since the dam was formed. At present Lake Sarez is 500 m deep, with the mean water level at 3,263 masl, with a minimum freeboard of about 50 m. The lake volume is 17 km³.

Despite the condition of the Usoy dam, a thriving community of over 28,000 people inhabit the narrow Murgab and Bartang valleys immediately downstream from the dam to the Afghan border, and further inhabitants living downstream from this point to where the valley broadens out to become the Amu Darya plane. All the inhabitants of these valleys, estimated to be around 100,000 people, are considered to be at direct risk from a potential failure of the Usoy dam.

*Figure 5.9* Usoy dam: general view of the right abutment showing the area of the original 1911 landslide.

*Figure 5.10* **Right Bank landslide.**

## 5.4.2   Risks

There are over 12 active landslides identified along the lake rims, the closest and the most dangerous being the "Right Bank Landslide" (RBL) (Figure 5.10) which is some 2 km upstream of the dam, with an average annual rate of displacement of approximately 10 cm/year

Although the dam has not been overtopped in the past, water has seeped out of the base of the dam, at downstream toe (see Figure 5.11) at a rate which approximately matches the rate of inflow, maintaining the lake at a relatively constant level. The flow averages about 45 m$^3$/s.

## 5.4.3   Lake Sarez risk mitigation project

### 5.4.3.1   Pre-Lake Sarez risk mitigation project investigations and studies

Natural dams are by no means rare phenomena but are generally short-lived: 50% of the 100 cases studied (Ref. Droz) have failed within 10 days of their formation and 90% within 1 year. However, Usoy dam survived for over 100 years.

Vulnerability of this area in the Pamirs to landslides was investigated by Russian scientists in the 1880s, some 30 years before Usoy Dam was formed. After the 1911 event, several expeditions by Russian geologists were sent to the area to examine the dam and the landslides in its vicinity. All the works stopped during the Soviet Revolution from 1917 to 1925 and from 1925 to 1940 most of the work was focused on the downstream seepages from the dam. In 1968 a risk from the RBL was recognised and till 1991 intensive research and monitoring by Russian Scientists were undertaken at

*Figure 5.11* Usoy dam – seepages at the downstream toe.

the RBL. During the period from 1991 to 1999, after the break-up of the Soviet Union, no work was done on the Usoy dam and Lake Sarez.

During 1999–2000, under International Decade for Natural Disaster Reduction (IDNDR) some monitoring was reinstated at Usoy Dam and the RBL.

Although the safety of the Usoy dam has been studied over many years, significant gaps and inconsistencies existed in the data available. The risk to the downstream population living in the Bartang and Pyanj river valleys was high.

As explained in Section 5.3.3.1 above, safety risks of the Usoy Dam and Lake Sarez have been addressed under WAEMP in early 2000.

### 5.4.3.2 Lake Sarez risk mitigation project

The Lake Sarez Risk Mitigation Project (LSRMP), conducted from 2000 to 2004, was jointly founded by the WB and the Swiss aid agency SECO, to reduce the risk related to the natural dam structure by implementing a monitoring and early warning system, training the population leaving downstream and outlining possible long term solutions which will minimise the probability of failure or the impact of a dangerous situation.[7]

The work under the LSRMP included:

1. Date collection
2. Site visit and inspection of dam, downstream area, landslides within the reservoir rim (Figure 5.12). The scale & remoteness of the "site" were overwhelming. The visit was done by a helicopter
3. undertaking studies: geological assessment of the dam and the RBL, seismic assessment (Probabilistic Seismic Hazard Assessment), evaluation of hydrological data, floods and the lake level evolution and filtration processes through the dam;
4. Hazard Identification – the following hazards have been studied: dam instability due to overtopping and external erosion, seismic instability, internal erosion
5. Risk Analysis – An "Event Tree" method was used to estimate the Usoy Dam probability of failure, which was estimated to be of $5 \times 10^{-4}$. It was estimated that with the implementation of Monitoring and Early Warning System the population at risk will reduce from 100,000 to 1,000
6. Design and installation of Monitoring System (MS) and Early Warning System (EWS):
   - monitoring system comprises:
     - lake level monitoring
     - strong motion accelerometers
     - downstream flow measurement
     - settlement cells
   - Data collected and analysed (compared with trigger values) at the Central Unit and transmitted to Dushanbe via satellite link. The Central Unit is located on the dam
   - Five levels of warnings can be issued from dam directly to population at risk by radio

*Figure 5.12* Site visit by a helicopter.

7. Preparation of Emergency Preparedness Plan – the plan defined the trigger levels and action required for different triggers
8. Identification of Potential Long-Term Structural Measures – the measures studies are (i) permanently lowering the lake level by tunnels that could be constructed at the left abutment, and (ii) raising the dam crest and strengthening the dam. Measures under (i) are regarded as the best in the long term structural measures.

On 24 January 2011, Magnitude 6.1 earthquake occurred within the Lake Sarez itself and it was picked up by the newly installed instruments.

## 5.5   CONCLUSIONS AND CHALLENGES

The safety of two unique dams, Nurek Dam and Usoy Natural dam was assessed under Dam Safety projects undertaken from the early 2000s until to date. Each dam is exceptional: Nurek dam is 300 m tall and is now the second tallest dam in the world, and Usoy dam is 500–700 m tall, and is the tallest natural dam in the world.

The safety of both dams is extremely important as many thousands of people live immediately downstream of the dams.

Comprehensive safety remedial works, which are currently being implemented at Nurek dam, will involve rehabilitation of the spillway tunnels, seismic belts, etc. The rehabilitation measures will also include enhancement of the monitoring instrumentation and review of EPP, O&M and the Instrumentation Plan. Once the rehabilitation works are implemented, the dam will satisfy international standards for dam safety; the EPP will ensure that the dam poses a reduced risk to the people living downstream.

Location, remoteness, and issues associated with the Usoy Natural dam are overwhelming. After assessing the dam's safety, there was a reasonably comfortable feeling about the condition of the Usoy Dam itself as it survived over 100 years while 90% of landslide dams failed in the first year. The immediate danger to the downstream population has been minimised by the installation of EWS, EPP, identification of safe havens and training of the population.

**Challenges**

Design Standards – As for Armenia and Georgia, Tajikistan also inherited dams which were designed, constructed, and monitored in accordance with SNIP norms at the time; these norms differ significantly from the International Standards, especially when adopting design and safety check floods, as well as design seismic acceleration.

Negligible Budget for Monitoring and Maintenance – this has been the case for both the Nurek dam and the Lake Sarez.

Political Aspects – Political situation in Tajikistan after independence in 1991 played a significant role in both dams' lifetime; Nurek dam was designed, constructed and initially operated by the Soviet engineers; Lake Sarez was investigated and monitored by the Soviet engineers. However, after the break up of the Soviet Union, both dams were left in hands of Tajik engineers who lacked a lot of original investigation and design data, as these were not made available to them.

Impacts of Dams' Failure – Social implication of both dams' failure would be catastrophic; in a case of a dam break, both Nurek dam and Lake Sarez would drain into the Pyanj River, which is a transboundary river. Therefore, the safety of both dams has been a priority for the Tajik Government.

**Positives**

Through these dam safety projects and the current Nurek HPP rehabilitation projects, extremely good collaboration and knowledge transfer have been achieved between international consultants and the Tajik engineers.

Operators of Nurek Dam, staff involved in the monitoring of Usoy Dam, as well as staff from various government organisations have been trained on various dam safety aspects, including management of emergencies.

## NOTES

1 Concerning the problem of seismic stability of the Nurek dam, Savinov, O.A., et al., Hydrotechnical Construction, 1986.
2 Induced seismicity at Nurek Reservoir, Simson, D.W., Negmatulaev, S.K., Bulletin of the Seismological Society of America, 1981.
3 Interstate Water Resource Risk Management Towards a Sustainable Future for the Aral Sea Basin, Olsson, O., et al., IWA Publishing, London, 2010.
4 Long term simulation of reservoir sedimentation with turbid underflows, Petkovšek, G., 2018.
5 Lake Sarez Risk Analysis, Attewill, L. and Spasic-Gril, L., Dams and Reservoirs, BDS, May 2001.
6 Sarez Risk Mitigation Project, Attewill, L. and Spasic-Gril, L., Lake 12th Biennial BDS Conference, Dublin, 2002.
7 Lake Sarez Risk Mitigation Project: A global risk analyses, Droz, P. and Spasic-Gril, L., IAHR Symposium, St. Petersburg, 2002.

# Chapter 6

# Case study

## South Asia – Sri Lanka

### 6.1  ABOUT THE COUNTRY

Sri Lanka is an island that lies on the Indian plate and lies in the Indian Ocean to the southwest of the Bay of Bengal. It is separated from the Indian subcontinent by the Gulf of Mannar and the Palk Strait. It has coastal plains, with mountains in the central area. The highest peak is Pidurutalagala at 2,524 m above sea level. It also has many rivers (103 in total), the longest of which is the Mahaweli River which is 335 km long. There are over 50 waterfalls, and over 40 estuaries and lagoons (Figure 6.1).

**No. of people** – 22.16 million
**Area** – 65,610 km$^2$
**Population Density** – 338/km$^2$
**Economy:**
GDP Nominal – (Total) $92.11 billion

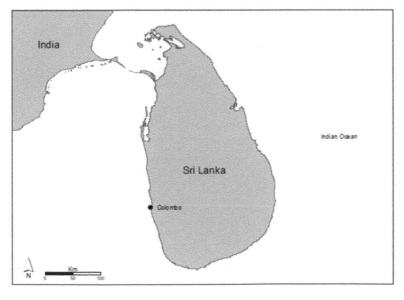

*Figure 6.1*  Map of Sri Lanka.

DOI: 10.1201/9780429320453-6

GNI Nominal – (per capita) $4,020

**Climate** – The climate is tropical and warm, and it varies because of the winds coming in off the ocean, the temperature ranges from 17°C to 33°C. There are monsoon seasons, and the rainfall is influenced by the monsoons, which come from the Bay of Bengal and the Indian Ocean. Some of the highlands have up to 2,500 mm of rain each year; however, most parts of Sri Lanka have about 1,200–1,900 mm of annual rainfall. It is also known to have tropical cyclones in humid and mountainous areas, which can cause flooding and problems to the infrastructure.

## 6.2   ABOUT THE DAMS

**No. of Dams** – There are over 300 dams in Sri Lanka, 34 dams are higher than 10 m, the tallest of which is the Victoria dam standing at 122 m (see Figure 6.2). Twenty-two dams are for hydropower, and the others are for irrigation.

The history of dam engineering in Sri Lanka dates to the 4th Century BC[1] when ancient Ceylon developed control of the water streams to satisfy the needs of an advanced civilisation. These great works of irrigation, some of which have been in use to this date, are impressive and attract even more interest, than many remains of ancient monuments, palaces, and temples.

One such reservoir is at Polonnaruwa in the northeast of Sri Lanka, called on account of its size the Parakrama Samudra (Sea of Parakrama). With an area of 30 km$^2$ and an enclosing embankment 14 km long, it irrigated an area of nearly 100 km$^2$. The dam is over 2,000 years old and is still operational to date (Figures 6.3 and 6.4).

Another remarkable example of an ancient dam is Tissa Wewa, shown in Figure 6.5 below, situated in Tissamaharama, in the South of Sri Lanka.

*Figure 6.2* **Victoria arch dam, Sri Lanka.**

*Figure 6.3* Polonnaruwa dam outlet structure, Sri Lanka over 2,000 years old.

*Figure 6.4* Polonnaruwa dam, Sri Lanka over 2,000 years old.

Tissa Wewa is a reservoir thought to have been constructed in the 3rd Century BC to irrigate paddy lands and supply water to the flourishing city of Tissamaharama. It is supported by an embankment. The lake was restored in 1871.

**Loss of Lives due to Dam failure**

Two dam failures, in 1986 and 2013, resulted in 127 people killed, affected 10,864 residents, destroyed 1,200 houses, and damaged agriculture.

Figure 6.5 **Tissa Wewa reservoir.**

**Law and regulations**

There is no Dam Safety Legislation in Sri Lanka. Dam Safety and Reservoir Conservation Project recommended the establishment of a Dam Safety Regulation; however, that could not be implemented so far (see Section 6.3 below).

**ICOLD membership**

Sri Lanka has been a member of ICOLD

## 6.3   DAM SAFETY AND RESERVOIR CONSERVATION PROJECT

### 6.3.1   Project background

The Dam Safety and Reservoir Conservation Project (DS&RCP) of Mahaweli Reconstruction & Rehabilitation Project (MRRP) funded by the International Development Agency and implemented from 2000 to 2004 had an objective to implement a qualitative management system for all major dams in Sri Lanka in order to improve their safety.

In 2000, under the MRRP a Risk Assessment study of the 32 major dams in the Mahaweli river basin and adjoining basins was conducted.[2] The study showed that while the modern dams have generally been built to current standards of the world's best-used practices, the same cannot be said for the other dams. Many dams are showing signs of ageing while others have significant deficiencies in monitoring, maintenance, reservoir conservation and other issues. A vast majority of dams have not had an overall safety review and risk assessment.

The main objective of the DS&RCP was to assess the safety of the selected 32 major dams and to recommend remedial works as well as to assist Sri Lanka in establishing a long-term dam safety programme.

## 6.3.2    Profile of dams

The DS&RCP covered 32 dams. All dams have been classified as Large Dams in accordance with the ICOLD classification. They can be categorised as follows:

**Mahaweli multipurpose dams**

Four of the 32 dams are large modern dams on the Mahaweli river serving both hydropower and irrigation purposes: Kotmale and Randeningala (rockfill), Victoria (arch) and Rantembe (concrete gravity). In addition, the Polgolla diversion barrage supplies the Sudu Ganga and associated power stations and irrigation schemes.

The five dams are owned and operated by the Mahaweli Authority of Sri Lanka (MASL).

**Hydropower dams**

Six of the 32 dams are single-purpose, hydropower dams, owned and operated by the Ceylon Electricity Board (CEB). Five of these dams are concrete gravity dams on the Laxapana river system constructed in the 1950s. The sixth, the Samanalawewa dam (Figure 6.6), is a rockfill dam constructed in the 1980s.

**Irrigation dams**

Most of the dams are single-purpose irrigation dams and are owned and operated either by the Irrigation Department (ID) of the Ministry of Agriculture or the MASL. Thirteen of the irrigation dams owned and operated by the ID were originally constructed over 1,500 years ago and are still in use after successive rehabilitation and reconstruction campaigns.

*Figure 6.6*  110 m high Samanalawewa HPP dam.

### 6.3.3   Inspections and condition of dams

All 32 dams were inspected early in the programme following a procedure typical for a periodic inspection under the UK Reservoirs Act 1975.[3] All the 14 dams owned and operated by MASL had previously been inspected, by the staff of the Sri Lankan consultancy, and reports were available. Irrigation dams are generally inspected monthly or quarterly by ID staff who complete a proforma report. There is no evidence of CEB dams having been previously inspected.

**Summary of condition**

Conclusion on the overall safety of the 32 dams from the work carried out under this activity is that there are very few unsafe dams, but that there is a range of issues that need to be addressed in order to preserve and in some instances to improve the status quo. Adequate dam safety depends on three separate factors: design, construction, and operation/maintenance.

Although the design of the dam's ranges from the simple homogenous embankments of the ancient dams to the sophisticated double curvature arch of Victoria, there is no instance where the safety of a dam is jeopardised by poor design.

There are several dams where the standard of construction has been below an acceptable level, and at several dam's poor constructions may jeopardise dam safety.

Generally, maintenance is barely adequate, and if this situation is not improved the safety of the dams will slowly deteriorate.

**Recommendations**

The following recommendations were made:

* Remedial works, categorised by priority
* maintenance items
* instrumentation and monitoring
* investigations and studies
* the nature, frequency, and scope of future

**Spillways**

*Spillway capacity*

Assessment of the adequacy of spillway capacity was comprised, for all 32 dams, the collection, review, and detailed analysis of all hydrological data relevant to the dams.

Two methods were used for estimation of the design inflow floods: the statistical approach which is based on historic records of the annual maximum flows recorded at all gauging stations in Sri Lanka and the unit hydrograph method. The statistical approach is based on the maximum annual flows for each year of record for the 80 gauging stations in Sri Lanka, providing some 2,000 station years of record.

Because of the high density of population downstream of the dams, spillway capacity was also checked for the PMF. The PMF inflow hydrographs were obtained by a simplified version of the unit hydrograph method and the estimation of the probable maximum precipitation (PMP) over the catchment.

The PMP was estimated from the maximum recorded rainfall at each meteorological station over the period of record, which for many stations exceeds 100 years.

The check of the adequacy of the spillways and other outlets of the 32 dams showed that all but three of the spillways had adequate capacity: for these dams, extra capacity can be economically and safely provided by heightening the dams concerned.

*Spillway reliability*

Of the 32 dams, 22 are either wholly or partly dependent on gated spillways for their safety. Of these spillways, 19 are electrically actuated, although most are capable – in theory – of manual operation. This is not a satisfactory safety condition and installation of stand by generators is recommended.

Nine of the 22 gated spillways were rated high reliability with no significant remedial works required.

**Stability**

*Embankment dams*

Among the 22 embankment dams, only the 4 modern rockfill dams and one zoned embankment had geotechnical information available from the original design stage which proved that the dams were stable. The geotechnical information for the remaining 17 earth embankments (13 of which are ancient) was either poor or non-existent. Therefore, the stability of these 17 dams was carried out using an assumed range of lower bound strength parameters.

Based on the stability results 17 dams were grouped into the following three groups:

- Group 1 – FOS < 1.3- Investigation required (four dams)
- Group 2 – 1.3 <FOS < 1.5- Investigation required if high groundwater levels or specific defects were identified in the inspections (seven dams)
- Group 3 – FOS > 1.5- No investigations required (six dams)

It was recommended that four dams from the first group site investigations be carried out and the stability reassessed using the parameters from the investigation. In addition, three other dams from the second group also required investigations because of defects identified during the dam inspections.

*Concrete dams*

Out of ten concrete dams, nine are gravity dams with heights varying from 18.3 to 42 m, and Victoria dam, a 122 m high concrete arch dam on the Mahaweli Ganga.

Safety of the dams to sliding and overturning as well as the stress at the key points was checked for the normal, unusual and extreme loading conditions.

Seven dams were found to be stable with an adequate safety margin under all loading conditions. However, three dams were found not to have sufficient safety margin and appropriate remedial works – improved foundation drainage - were recommended.

**Instrumentation**

It was found during our inspection that the dams constructed recently were equipped with electronic instrumentation to measure seepage, pore pressure, deformations, deflections, movements, temperature, and various other parameters.

This equipment, whilst operating well for several years, has rarely been serviced or calibrated. Where equipment has failed there has been little funding available for its maintenance or repair which has resulted in the equipment being abandoned. In some cases, a lack of understanding of a system has led to the equipment being abandoned or deemed inappropriate.

The dams that were constructed in the mid-20th century have fewer instruments, and the ancient dams usually have no instrumentation at all.

Currently, dam monitoring is undertaken by dam owners and on many of the sites, the monitoring is carried out regularly. However, data recording and handling procedures often vary from site to site. The instrument monitoring staff has a basic understanding of the instrument operation, but the data handling procedures are not standardised.

Following the inspection, we have recommended and specified additional instruments: these comprise for most dams the collection and measurement of seepage and the provision of survey monuments to enable settlement surveys to be carried out.

Standardisation of data recording and presentation was proposed. It was also proposed that the records will be in a centralised data record library within the Data Management Centre in Colombo and will be available via the GIS system.

### 6.3.4 Operation and maintenance (O&M)

The perceived shortcomings in present O&M procedures are as much the product of inadequate budgets and the failure of management to recruit, train and financially reward staff of the calibre necessary to operate and maintain large dams, as they are deficiencies in management procedures and practices.

This in turn may be seen as being a failure by the Government to recognise the importance of the security of the nation's stock of large dams to the national economy, and the threat that unsafe dams pose to the public at large.

For this reason, it has been necessary to take full recognition of the initiatives that have been discussed to restructure the main water management agencies, to introduce a new Water Act and to set up a regulatory framework for dam safety.

The form that the regulatory framework will take will impose obligations on dam owners that will significantly affect the procedures to be adopted for O&M and safety surveillance. Prior to the preparation of Guidelines for Improvement of O&M and Emergency Procedures we examined the current practices which are applied within each of the agencies. They are summarised below.

*   MASL – Procedures for O&M of the large MASL dams are now well established. All of the new dams have O&M manuals prepared by the designers which set out routine procedures for O&M as well as emergency procedures, particularly in the event of a major flood.
*   CEB – Procedures for the operation of CEB dams are determined in Colombo to meet energy requirements within the distribution system. The procedure adopted is that gate operating staff are assigned to provide 24-hour cover at each of these dams whenever the water level approaches FSL and continues until the water level has again fallen below FSL.
*   ID – Operation of the ID dams is regulated by a departmental circular which covers the whole irrigation scheme as well as the headworks.

### 6.3.5 Emergency preparedness

Some effort has been made at the big dams to prepare for emergencies, in that key staff have been listed with their home contact details, contact details have been compiled for the emergency services and other key authorities, and lists of emergency service providers have been made.

But generally, there has been no attempt to identify risks, to set levels of alarm in response to different emergencies, or to determine the actions and persons responsible in any set of circumstances. Also, there is no programme of formal training for operating staff in dealing with emergencies.

**Prepared Guidelines for Improved O&M and EPP**

It was proposed under DS&RCP that improved management practice for Sri Lanka's stock of large dams required that the three principal agencies adopted a structured, simple, and standardised approach to O&M and Emergency Preparedness.

The guidelines were drawn up for the preparation of Standard Operating Procedures (SOPs) and EPPs for all dams in Sri Lanka. Prototype documents were also produced that are intended for universal application by the three agencies.

## 6.3.6   Reservoir conservation

**Extent of sedimentation and pollution**

In broad terms, Sri Lankan reservoirs are not severely affected by either sedimentation or pollution. However, the pressures exerted by a rapidly expanding population have resulted in environmental degradation of one-third of the total land area.

Soil erosion is most severe in the high catchments on steep slopes at mid-levels, which are used for market gardening and tobacco production: it is estimated that erosion rates for these land uses are 150 t/ha/year, compared with 0–10 t/ha/year for paddy, forest, and well-managed tea.

The actual sediment yield of the catchments varies between 0.5 and 4 t/ha/year for lowland and upland reservoirs, respectively. Of the 32 reservoirs studied only two, Polgolla and Rantembe are seriously affected by sedimentation.

Similarly, water quality is becoming a more serious problem because of increasing levels of nutrients, pesticides and effluents entering the watercourses.

**Conservation policy**

A national conservation policy is required to reverse the adverse trends in sedimentation and water quality to protect the country's water resources. Sediment yields will be reduced, and the water quality improved by:

• propagation of appropriate land use, including grassing or reforesting steep and high-level areas currently used for agriculture, the prevention of overgrazing and the adoption of soil conservation measures
• the adoption of appropriate land use and fiscal policies to improve land tenure systems and discourage the fragmentation of land
• improvement of urban wastewater treatment and the disposal of solid waste
• better management of fertilisers and pesticides
• enforcement of the 100 m buffer zone of grassland and trees around the reservoir perimeter.

Considerable efforts are already being made in the conservation of the Mahaweli catchments, including research, public awareness, and farmer training. This work needs to be intensified and extended to all catchments.

### 6.3.7   Dam safety legislation

Under DS&RCP a paper outlining the main provisions of future Dam Safety Legislation in Sri Lanka, based on a review of legislation in the UK, the USA, Sweden, and India was prepared. The main provisions of the proposed legislation were:

*   The dam owner is responsible for the safety of the dam
*   A register of dams would be compiled and maintained by the enforcement authority
*   Dams would be subject to mandatory inspections by independent engineers
*   Recommended remedial works would be mandatory

After much internal discussion, the Client decided that Sri Lanka was not ready for legislation and that the proposed provisions should be contained in a Code of Practice. The DSMC will be responsible for monitoring compliance with this Code. This is still in place to date.

### 6.3.8   Portfolio risk assessment

PRA has been undertaken to provide a rational method of improving the safety of the dams.

The risk for all 32 dams was assessed by the semi-quantitative method, described in Sections 3.3 and 4.3. The results of the risk analysis are shown on the F-N plot in Figure 6.7. It could be seen that by the implementation of rehabilitation works and implementation of Emergency Preparedness Plans probability of failure of several dams will be reduced. These measures will provide a significant reduction in the potential loss of lives.

**Dam Safety rehabilitation programme**

It has been recommended that the dam safety rehabilitation programme comprised implementation of both structural and non-structural measures, as follows:

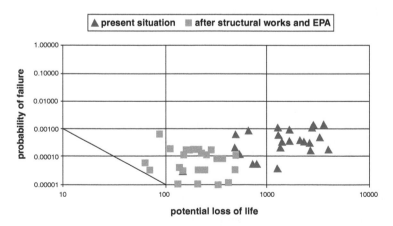

*Figure 6.7* Potential loss of life vs probability of failure.

**Structural measures**
Improvements to spillways and outlets
Repairs to upstream slope protection
Dam and foundation drainage
**Non-structural measures**
Installation of Monitoring Systems
Installation of Early Warning Systems

## 6.4   CONCLUSIONS AND CHALLENGES

Dam Safety rehabilitation measures are recommended to be implemented under DS&RCP on 32 dams comprised rehabilitation works, including structural measures and implementation of Monitoring Systems, EWS and EPPs. It was estimated that these measures would reduce the probability of failure of several dams and provide a significant reduction in the potential loss of lives.

The CEB dams that were assessed under this portfolio had not been previously inspected and no inspection records were available.

O&M – The perceived shortcomings in present O&M procedures are as much the product of inadequate budgets and the failure of management to recruit, train and financially reward staff of the calibre necessary to operate and maintain large dams, as they are deficiencies in management procedures and practices. This may be seen as being a failure by the Government to recognise the importance of the security of the nation's stock of large dams to the national economy, and the threat that unsafe dams pose to the public at large. These aspects need to be addressed by the government.

Dam Safety Legislation – After many internal discussions under DS&RCP, the Client decided that Sri Lanka is not ready for legislation and that the proposed provisions should be contained in a Code of Practice. The DSMC will be responsible for monitoring compliance with this Code. This is still in place to date. It would be recommendable for Sri Lanka to consider the development of the Dam Safety legislation.

## NOTES

1  Sri Lanka Ancient irrigation, Parker, 1881.
2  Sri Lanka Dam Safety and Reservoir Conservation Programme, Attewill, L., Spasic-Gril, L., and Penman, J., 13th Biennial BDS Conference, Canterbury, 2004.
3  UK Reservoirs Act 1975.

Chapter 7

# Case studies

## South East Asia – Myanmar

### 7.1  ABOUT THE COUNTRY

Myanmar, shown in Figure 7.1, is bordered in the northwest by Bangladesh and India, in the east by China, Laos and Thailand. It has 1,200 miles of continuous coastline along the Bay of Bengal in the southwest. In the north, the Hengduan mountains form its border with China, the tallest of which stands at 5,881 masl - Hkakabo Razi. The mountains divide the three main rivers - the Irrawaddy, Salween and Sittaung. The Irrawaddy (Figure 7.2) is Myanmar's longest river and is 2,288 km long. The mountain chains contain valleys with fertile plains; most of the population live in the Irrawaddy valley.

**No. of people** – 53.6 million
**Area** –676,578 km$^2$

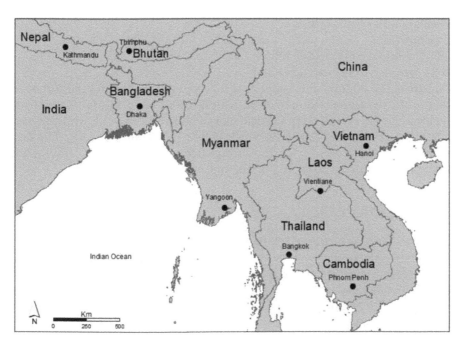

*Figure 7.1* Map of Myanmar.

DOI: 10.1201/9780429320453-7

*Figure 7.2* Irrawaddy river in Mandalay.

**Population Density** – 76/km$^2$
**Economy:**
GDP Nominal – (Total) $66 billion
GNI Nominal – (per capita) $1,390
   **Climate -** The country lies between the Tropic of Cancer and the Equator, and is firmly in the monsoon region of Asia, with a 5,000 mm of rainfall annually in the coastal areas; however it is less in the delta region where it is halved, whilst the average rainfall in the dry zone of the country is less than 1,000 mm annually. The north of Myanmar is the coolest area, where the average temperatures are 21°C; but the average maximum temperatures rise to 32°C in the coastal and delta regions.

## 7.2   ABOUT THE DAMS

**No. of Dams**: There are almost 200 dams in Myanmar, mostly for hydropower. The number of dams has tripled over the last 30 years, with the Government programme to provide water for irrigation and hydropower.
   **Law and regulations**
   The Myanmar National Committee for Large Dams enacted Dam Law in 2015 (see Section 7.4 below).
   **ICOLD membership**
   Myanmar has been a member of ICOLD

## 7.3   DAM SAFETY ASSESSMENT OF FIVE LARGE IRRIGATION DAMS

### 7.3.1   Background

As part of the preparation of the Agricultural Development Support Project (ADSP), which has an activity to reinforce the capacities of existing dams for water supply, a dam safety assessment has been undertaken in 2014 for five existing irrigation dams owned by the Ministry of Agriculture and Irrigation (MoAI).[i]

The dams are classified as Large Dam under ICOLD classification. The potential hazard level for these dams is generally Extreme, with large farming areas and communities downstream of the dams.

At the time the safety assessment was undertaken, the dams have been in operation between 7 and 15 years.

The following dams have been assessed:

- Dam A – Mandalay region
- Dams B & C – Sagaing Region
- Dam D – Nay Pyi Taw Region
- Dam E – Bago-East Region

Scope of the dam safety assignment included:

- Review of as-built or design drawings and construction specifications;
- Review of reports on geological investigations, materials testing, strength parameters and stability analysis;
- Review the criteria, methodology and determination of the design flood, flood routing studies and the spillway sizing. Examination of spillway operation records and evaluation of the adequacy of the spillway capacity;
- Review previous dam safety inspection reports as well as recent O&M records, and any instrumentation data, such as reservoir level, inflow & outflow volume, spillway discharge volume, seepage volume, settlement, etc;
- Perform field inspection and categorise the status as satisfactory, fair, poor, or unsatisfactory.
- Provide recommendations of required remedial and upgrading measures with provisional cost estimates;
- Assess the current operational procedures;
- Evaluate the organisational structure, staffing, budget, equipment and facilities needed to operate and maintain the dam in a safe and sustainable manner.

### 7.3.2   Main features of the dams

The main design features of the five dams are summarised in Table 7.1 below.

The dams comprise:

- An earthfill embankment;
- Irrigation intake and outlet works;

---

i  Since 2016 the MoAI has been changed to MoALI, Ministry of Agriculture, Livestock, and Irrigation

*Table 7.1* Main features of dam projects

| Item | Unit | Dam A | Dam B | Dam C | Dam E | Dam F |
|---|---|---|---|---|---|---|
| Year of completion | | 2007/8 | 2007 | 1998/9 | 1999/2000 | 2004/5 |
| Catchment area | km² | 192 | 66 | 75 | 798 | 1,044 |
| Annual rainfall (adopted in design) | mm | 1,090 | 946 | 850 | 857 | 2,127 |
| Dam height | m | 21.6 | 42.3 | 19.5 | 33.2 | 29.6 |
| Total storage capacity | MCM | 71 | 152 | 17 | 176 | 267 |
| Dead storage capacity | MCM | 9.1 | 26 | 4 | 21.35 | 35 |
| Dead storage/total storage capacity | % | 13 | 17 | 20 | 12 | 13 |
| Design flood return period | | 1 in 1,000 | 1 in 1,000 | 1 in 1,000 | 1 in 1,000 | 1 in 1,000 |

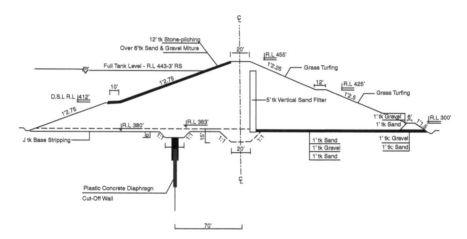

*Figure 7.3* Typical cross section of embankments.

- Surface spillway
- Some dams have intake, penstock and small hydropower plant (HPP)

All dams have a similar cross section, which is shown in Figure 7.3; it typically includes a rip-rap upstream slope protection, a cut-off-trench in the foundations, a diaphragm wall in the foundations (partially), a vertical chimney filter, which leads into a horizontal drainage blanket and a drainage ditch. The embankment fill materials are clayey and sandy silts.

### 7.3.3  Issues arising from site visits and dam safety reviews

*7.3.3.1  Hydrology, hydraulics and sedimentation*

**Hydrological data**

Use of unconservative and underestimated hydrological data was common for all five dams. This led to significantly underestimated inflows.

For Dam A, which was completed in 2007/8, a high hydrological variability was noted: it appeared that in 2010, 2011 and 2013 the annual inflows were almost twice the storage capacity and the surplus inflows had to be spilled.

Dam B, completed in 2008, was in 2011 exposed to the $O_2B$ Cyclone; during the Cyclone the water level in the reservoir was almost at the dam crest, which was subsequently raised by 1.5 m. The heightened portion is supported by L-shaped retaining walls as shown in Figure 7.4. The spillway chute slab was washed out during the 2011 cyclone and was subsequently repaired.

For Dam C, completed in 2009, between 2003 and 2010 an average annual inflow was over 20 times more than the storage available; all the surplus inflows had to be spilled via the service and emergency spillways.

For Dam D, monthly storage balance records from 1999 to 2013 have been reviewed; it was noted that the annual inflow volume showed very high variation.

For dam E, monthly storage balance records from 2004 to 2013 were reviewed. It was noted that the inflow could easily fill up the reservoir every year and the spillway had to be used quite frequently for overspills.

**Spillways**

Spillways were designed for a Safety Check Flood of 1 in 1,000 years – when compared with the current practice for selection of the Safety Check Floods for Extreme Hazard Dams stated in the ICOLD Bulletin 82 (Endnote 10 in Chapter 2), or the FEMA[1] this flood is considered to be too low bearing in mind the height of the dam, its volume and population leaving downstream.

It has been noted during the site visit that stop logs are placed on the spillway crest (Dam A) and are used to raise the FSL by 1.5 m. This is regarded as unsafe as it would reduce the spillway capacity and raise the water level higher than designed. It would

*Figure 7.4* Dam B was raised by 1.5 m to prevent overtopping.

also be challenging to remove stop logs in a timely manner when the water level rapidly goes up during floods.

For Dam C, there is a serious concern with regards to safety of the dam and the ability of spillways to pass extreme floods due to significant sedimentation of the reservoir (see below).

For Dam E, the spillway chute was excavated in highly erodible natural soils and there are evidences of erosion and instability of the spillway slopes, which could affect the operation of the spillway (Figure 7.5).

### Sedimentation

For Dam A, sediment yield was assumed to be 1,100 $t/km^2$/year; however, this was not supported by measurements.

For Dam B, the design was based on an assumed sediment yield of 1,760 $t/km^2$/year, which was not supported by measurements. Based on the sediment yield measured 4 years after commission, the yield was around 3,000 $t/km^2$/year. It is recommended that sediment management should be considered to avoid heavy siltation of the reservoir storage.

For Dam C the design was based on an assumed sediment yield of 1,100 $t/km^2$/year. However, heavy sedimentation has been observed and a bathymetric survey was undertaken in 2003 (4 years after commission). The sediment yield measured was app. 3,000 $t/km^2$/year, which is categorised as heavy sedimentation. Based on this data, it could be seen that all the dead storage has been filled with sediments in less than 5 years of operation; the sediments could soon impact the operation of inlet gates for the main canal; for this reason, two additional canals have been constructed, with high-level off-takes, namely the right bank canal and the left bank canal.

*Figure 7.5* Dam E- erosion of slopes to the spillway chute.

For Dam D, no information on the reservoir sedimentation could be obtained, although the spindle of one of two intake gates was bent and needed to be repaired.

No information on the reservoir sedimentation could be provided for Dam E.

### 7.3.3.2    Embankment design

In the original design, all embankments had the same, standard cross section, and the foundation improvement measures, regardless of the ground conditions. This is not the most sustainable design, especially when considering the foundation seepages and measures to mitigate them (see 7.3.3.2.2 below). Also, dispersive characteristics of the embankment fill material have not been adequately investigated, and the design could be considered safe against the dispersive properties of the fill (see 7.3.3.2.3 below).

#### 7.3.3.2.1    SLOPE STABILITY

In the design, all embankments satisfy the required static stability factors of safety.

Slope stability in seismic conditions was not checked. Also, the slope stability calculation assumed that the downstream shoulder fill would be fully drained, i.e. no increase in phreatic surface were checked.

#### 7.3.3.2.2    SEEPAGE THROUGH THE FOUNDATION

Dams A, D and E have serious issues with foundation seepage and uplift at the downstream toe. These are described below:

**Dam A**

Seepage through the more permeable foundation soils was recognised as a risk during the design stage when the installation of a 20–25 m deep diaphragm wall was envisaged in the central embankment section. Between the Chainage 2,800 and 5,200. It is understood that a diaphragm wall was installed in this deepest section.

However, the seepage issues and potential problems with the uplift pressures underneath the downstream toe of the embankment in the right flank area from chainage 5,200 to 11,850 must have been recognised from the design. It was understood that it was originally planned to install an impervious blanket on the upstream toe of the dam, at the reservoir floor, over the entire length of the dam to alleviate potential seepages and uplifts at the downstream toe. The blanket might have subsequently only been installed over a short length in the middle of the dam.

Toward the end of construction in 2007 pressure relief wells were installed near the downstream toe, in the right flank, from Chainage 5,200 to 11,850; 26 relief wells were installed in 2007 and 35 new wells were added in 2011. However, during the time of the dam inspection in 2014, most of the wells became clogged and dysfunctional due to the absence of periodic cleaning/flushing works. There are no piezometers installed in the area with the relief wells that would measure the uplift pressures and efficiency of the wells.

During the site visit, several areas around Chainage 7,600–9,000 have been found to have very high leakages, as well as several sand boils at the downstream toe (Figure 7.6 below).

*Figure 7.6* Dam A- sand boils at the downstream toe.

In a summary, there are two issues related to the seepage from Chainage 5,200 to 10,000, namely:

• The excessive water losses were measured and observed during the site visit; and
• Possible risk for development of high uplift pressures at the downstream toe, at the bottom of the thin, less permeable layer of silt that covers a layer of very permeable gravel.

**Dam D**

The design report shows that a sheet pile wall or a grout curtain was envisaged to be installed in the foundation as anti-seepage measures along the dam axis. However, it is not known if these measures were constructed. During site visit, some marks of sand boils at the downstream toe area of the dam were noticed. These may be signs of high uplift pressures developing at the downstream toe, which should be carefully monitored and investigated when the water level is higher.

**Dam E**

As for Dam D, the design report shows that a diaphragm wall was envisaged to be installed in the foundation as anti-seepage measures along the dam axis. However, the actual extent of the wall constructed is not known. During the visit, some marks of sand boils at the downstream toe area of the dam were noticed (see Figure 7.7). These may be signs of high uplift pressures developing at the downstream toe, which should be carefully monitored and investigated when the water level is higher.

7.3.3.2.3   EMBANKMENT FILL MATERIAL PRONE TO DISPERSION

The embankments were typically constructed of clayey, sandy silts; the gradings typically show 1%–8% of clay, 50%–60% of silt and 35%–40% of sand. Such materials, the low to medium plasticity silts, could be prone to dispersion when exposed to water.

*Figure 7.7* **Dam E - downstream slope erosion.**

Endnote 2 states that several tests can check material's dispersion, such as the Crumb Test, Pinhole test etc; during construction, the Crumb Test is performed for acceptability of the borrowed materials. Suitability is defined as the Emerson Class Number (ECN). The soils are graded according to class, with Class 1 being the highly dispersive, Class 8 non-dispersive.[2] states that the soils with ECN Class 1–4 need to be treated with extra caution in dam construction.

Sherard et al.[3] indicate that if a soil tests as dispersive in the Crumb Test, it will also test as dispersive in the Pinhole test, but 40% of soils testing as dispersive in the Pinhole test, test as non-dispersive in the Crumb Test. Therefore, the Crumb Test is normally regarded as a useful first check on dispersivity, and further Pinhole tests are always recommended.

During construction of the embankments, the fill material was tested for dispersivity by the Crumb Tests only; most of the results of the testing show ECN Class 5–6, with some results as low as 1 and 2. It is therefore reasonable to assume that there might be areas within the embankment where the fill material with ECN < 5 was used for the embankment construction.

The embankment fill in A, C, D & F dams has been found to be susceptible to dispersion.

Figure 7.8 below shows large holes on the downstream slope of embankment D, which are typical for dispersive material; the Crumb Tests carried out during construction gave ESNs between 5 and 6.

It is recommended to take samples from the fill material in the downstream slope of embankments A, C, D and E and perform the Pinhole tests, and if required, implement remediation measures against soil dispersion.

### 7.3.3.3   Monitoring instruments

No piezometers were installed within the embankment which would confirm the position of the phreatic surface within the body of the embankment. Also, no other

*Figure 7.8* Dam D - large erosion holes on the downstream slope.

monitoring instruments exist. Only seepage monitoring weirs were installed on the downstream toe of the embankments.

### 7.3.3.4   Dam inspection reports

The dam inspection is undertaken once a week; however, as there are no monitoring instruments, except the seepage weirs, most of the observations that are made during inspections are only visual.

### 7.3.3.5   Safety equipment, operation & maintenance (O&M) and emergency preparedness plan (EPP)

Staff from the MoAI that operate the dam are based near the dam site; no safety equipment is present at the site and the communication with the Head Office is via a radio link. There are no O&M and EPP.

### 7.3.3.6   Access to dams

Access to some dams was difficult after heavy rains when they could only be accessed via motorbikes.

For Dam D, the access to the dam is impassable after heavy rains (see Figure 7.9), which would make emergency access to the dam after rain impossible by vehicles. This is not acceptable from a safety point of view.

*Figure 7.9* – Dam D - access road is impassable after heavy rain.

### 7.3.3.7    *Trees and bushes at the downstream slope and toes area*

On several dams, trees and bushes at the downstream slope and toe area of the dams have been encountered during site inspections (Figure 7.10); these could obstruct areas where the water is leaking.

Also, the paddy irrigation encroachment is sometimes too close to the downstream toe to enable leakage/turbidity monitoring (Figure 7.11).

## 7.3.4    Recommendations in the interest of safety

### 7.3.4.1    *Specific recommendations for the five dams visited*

1. Hydrological Data – Review hydrological data used in the design and adjust flood inflows; where necessary install advanced hydro-met monitoring & system to improve hydrological data
2. Geological Investigation – Undertake supplementary geological investigation for Dams A, D and E to understand geological conditions where the sand boils occur and design adequate remedial measures,
3. Sedimentation – Undertake a bathymetric survey to assess the actual sediment yield; check the capacity of the live storages and, if necessary, implement sedimentation mitigation and catchment sediment management measures in order to prolong the lifespan of the dams,

*Figure 7.10* Dam C- bushes at the downstream toe.

*Figure 7.11* Dam A – paddy irrigation is too close to the downstream toe to enable leak-age monitoring.

4. Dispersivity – for dam where the dispersive fill was used for embankments, map the areas on the d/s slope that show high erosion, sample and test for dispersivity by performing the Pinhole Test. If the soil is dispersive, perform mitigation measures (see 6 (iv) below),

5. Design Checks – (i) estimated risk-based design/check floods, (ii) check spillway capacity and available freeboard for floods >1 in 1,000 taking into account actual live reservoir storage; (iii) undertake seepage and stability analysis under increased uplift pressure condition; (iv) check seismic stability for an OBE and SEE; (v) check the total settlements versus the current position of the chimney crest; (vi) check the adequacy of the slope surface drainage channels to collect and discharge the surface run-off,

6. Remedial Measures – (i) where necessary, provide measures that will lower uplift pressure in the downstream shoulder, (ii) provide the toe drainage, (iii) repair gulley erosion and surface drainage channels, (iv) if the soil is dispersive (see 4 above), improve embankment surface protection by implementing soil stabilisation using lime, gypsum or aluminum sulphate or placement of non dispersive clay, (v) undertake spillway repair,

7. Instrumentation – prepare an instrumentation plan & install piezometers within the embankment body and at the downstream toe for measuring pressures in the relief wells and the areas adjacent to the wells, install benchmarks for measuring settlements,

8. Prepare an O&M Plan and EPP for all dams,

9. Monitoring – (i) carry out continuous monitoring of leakage volume and reservoir level, (ii) perform turbidity checking, (iii) check phreatic surface by piezometers,

10. Dam Safety Equipment – prepare a stockpile of sandbags and plastic sheets at the dam site ready to be placed around springs in order to counteract possible water boiling,

11. Ensure access to the dam under any weather conditions,

12. Others – clear the trees and bushes at the downstream slope and toe area of the dams and limit the paddy irrigation encroachment for leakage/turbidity monitoring.

### 7.3.4.2    High-level recommendations

Based on discussions and observations made during the site visits, it is recommended that the relevant ID's staff undertakes training in:

- Principles of Sustainable Dam Design, Construction, and Operation – it is important that certain aspects, are appropriately addressed during the design, construction and operation stages, as they may save the money in the long term by reducing expenditures allocated for more substantial maintenance, a need for dam safety modifications and upgrades or decommission of a dam.
- Potential Modes of Failure Analysis that will include hazard identification and risk assessment,
- Capacity strengthening and training on Dam Safety Plans.

In addition to the above, examples of certain aspects of a sustainable design, which address a long term hydrological assessment, selection of an appropriate design flood, sedimentation management etc, could also be presented during the training sessions.

It is also recommended that the following are considered/initiated in Myanmar by the ID management, in conjunction with other dam owners:

- Establishment of the National dam safety framework/legislation
- Preparation of Dams Inventory, registration and information database (around 300 dams)
- Preparation of Dam safety regulations and guidelines adapted to Myanmar including hydrological and seismic assessment

It is understood that the Government followed the recommendations related to the National dam safety legislation and the Myanmar National Committee for Large Dams enacted Dam Law in 2015.

## 7.4  CONCLUSIONS AND CHALLENGES

A rapid increase in design and construction of dams in the last three decades has put a pressure on dam designers and contractors to deliver projects quickly and economically; all the dams assessed had underestimated floods and sediment yield, and they all had the same typical cross sections with a lack of appropriate anti-seepage measures and absent monitoring instruments. Also, the ground investigation that was undertaken prior to or during construction was not adequate to detect geotechnical issues. These gaps in design and construction led to severe dam safety risks and a need for mitigation measures.

In addition to the political pressure, the above gaps in dam safety highlighted issues with sustainable design knowledge; these were discussed with the client, who was open to staff training and capacity strengthening. Also, a positive outcome of this project and discussions with the client was the adoption of the Dam Law in 2015.

However, a serious challenge that remains is dams' accessibility in remote areas, especially after floods, which needs serious improvement.

## NOTES

1  FEMA P-94: Selecting and Accommodating Inflow Design Floods for Dams, August 2013; US Department of Homeland and Security.
2  Geotechnical Engineering of Dams, by Robin Fell, Patrick MacGregor, David Stapledon, Graeme Bell.
3  Identification and Nature of Dispersive Soils, Sherard, J. L., Dunnigan, L. P., and Decker, R. S., 1976.

# Chapter 8

# Case studies

## Southeast Asia – Vietnam

### 8.1 ABOUT THE COUNTRY

Vietnam is located in mainland Southeast Asia (see Figure 8.1), on the eastern Indo-chinese peninsula; it is a very narrow, long country as the combined length of the land is 4,639 km. At its narrowest point, the country is 50 km across, but it widens to 600 km in the north. The terrain is mostly hilly and forested and mountains account for 40% of the country's land area.

The Red River Delta in the north of the country is a more developed and densely populated area than the Mekong River Delta in the south. Southern Vietnam has lowlands along the coast with mountains and forests. The highland account for around 20% of the arable land and 25% of the forested land. In the south, the soil is less fertile as there are relatively low nutrients because of cultivation. The highest peak in Vietnam is Fansipan, standing at 3,143 m. Its largest Lake the Ba Be Lake and the largest river is the Mekong River.

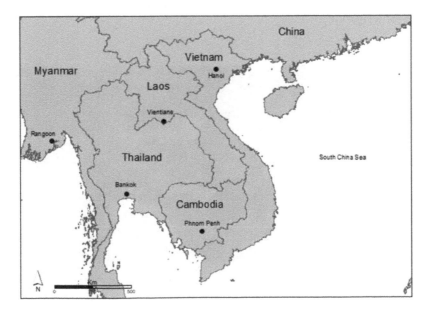

*Figure 8.1* Map of Vietnam.

DOI: 10.1201/9780429320453-8

The country is also prone to earthquakes which mainly occur in the provinces bordering Laos.

**No. of people** -96 million

**Area** – 331,212 km$^2$

**Population Density** – 295/km$^2$

**Economy:**

GDP Nominal - (Total) $261 billion

GNI Nominal - (per capita) $2,540

**Climate -** The climate tends to vary dramatically in Vietnam, considering what region you are in; due to the difference in latitude and topographical relief. The average annual rainfall is from 1,500 to 2,000 mm during the monsoon seasons which can cause flooding as the country is regularly affected by tropical storms and typhoons (see below).

**Vulnerability to climate change**

Vietnam is one of the most vulnerable countries to climate change, with 55% of the population living in low-elevation coastal areas exposed to the sea level rise.[1]

Also, the country is exposed to more frequent tropical storms; the storm in October 2020 had a death toll of 88 people. During the recent storm in November 2021, which affected central areas of the country, some areas had a rainfall of up to 600 mm within 24 hours. It was reported that over 7,000 people were evacuated to safety from flooded areas and areas at risk of landslides (Figures 8.2 and 8.3).

## 8.2   ABOUT DAMS

**Total no. of Dams (irrigation/hydropower) -** Vietnam has one of the largest dam systems in the world. The dam network comprises around 7,500 dams, with 640 medium and large dams and thousands of small dams (see Table 8.1 below).[2] About 96% of all the dams are for irrigation.

*Figure 8.2* **October 2020 floods.**

*Figure 8.3* November 2021 storm.

*Table 8.1* Classification of dams in Vietnam

| Dam height (m) | Reservoir volume MCM | Number of irrigation dams | Number of hydropower dams |
|---|---|---|---|
| >50 | | 3 | 32 |
| 15–50 | >3 | 551 | 54 |
| <15 | <3 | 6,648 | 201 |

According to Endnote 2, at present, the country has no national record of either small dams or their problems, for example, location, dam type and size, hazard rating, condition etc. There has been no systematic collection of data on dam failures and there have been no specific approaches to determine associated impacts or economic losses of failure. Ad hoc evidence suggests that many dam safety problems and notable dam failures have occurred in various provinces in Vietnam but have often been unreported. These failures have taken hundreds of lives and have caused devastating impacts on property and the environment. Even from just this limited recorded information, it is apparent that the costs of dam failures, including the associated threats to the security of agricultural produce

**Law and regulations**

Vietnam has developed laws and regulations on dam safety, but their implementation has not been effective.[3] It has been recognised that policy tools to drive better practice and establish levels of standards for operating of dams are required.

**ICOLD membership**

Vietnam has been a member of ICOLD

## 8.3 DAM REHABILITATION AND SAFETY IMPROVEMENT PROJECT

### 8.3.1 Project background

The Government of Vietnam has established a sectorial programme for dam safety in recognition of the importance of securing the foundations for sustained and secure economic growth, the Dam Rehabilitation and Safety Improvement Project (DRSIP). The DRSIP was first launched in 2003, revised in 2009 and again revised in 2015, as part of the effort to revitalise the programme activities and targets. Based on information available from the Ministry of Agriculture and Rural Development (MARD), there are about 1,150 irrigation dams in need of urgent rehabilitation or upgrading until 2022. The DRSIP is funded by the WB (93.3%) and co-funded (6.7%) by counterparts' funds.

The DRSIP is designed to improve the safety of the dams and related works, as well as the safety of people and the socio-economic infrastructure of the downstream communities. The project will also support the Government to ensure a more holistic, basin-level integrated development planning to improve institutional coordination, future development, and operational safety. The DRSIP objectives are to provide a mix of both structural and non-structural measures to selected dams.

Structural measures have been initially proposed for 442 dams in 34 provinces, see Figure 8.4 below. These dams were identified through an iterative, consultative prioritisation process with the national authorities and provincial agencies. Most of those dams are classified as small dams, with 71% being less than 15 m in height and with storage less than 3 MCM. There is a total of 104 dams higher than 15 m and 39 dams with storage of more than 3 MCM. Most of the dams for which data exists were constructed more than 15 years ago, with 50% constructed between 1970 and 1990.

Non-structural interventions have been proposed to support a range of national institutional and regulatory measures, as well as pilot specific basin level measures. These basin level measures are aimed at integrating dam and reservoir operations, improving data collection and information management within the basin context and facilitating specific coordination and governance mechanisms between sectors within the provinces as well as between provinces to introduce a more holistic, long-term approach.

All rehabilitation works have been designed by local consultants and implemented by local contractors.

The author has been involved in the DRSIP since February 2017 as the Chair of the International Dam Safety Panel of Experts. The Panel has been tasked to review safety conditions of some existing dams as well as the rehabilitation works proposed and undertaken for the most complicated dams. Over the period from 2017 to 2020 inspection and review of dam safety rehabilitation works of 19 dams have been undertaken by the Panel; 13 dams have been classified as Large and 6 dams as Small.[4] All 19 dams are homogeneous embankments. Many of the rehabilitation works that were proposed have been under implementation.

One of the Large dams is Dau Tieng Dam in the Tay Ninh Province, which is classified as a Special Class Dam in accordance with the Vietnamese Standards; the dam is the largest irrigation dam in Vietnam, it supplies water to more than *100,000 ha of*

VIETNAM: Dam Rehabilitation and Safety Improvement Project

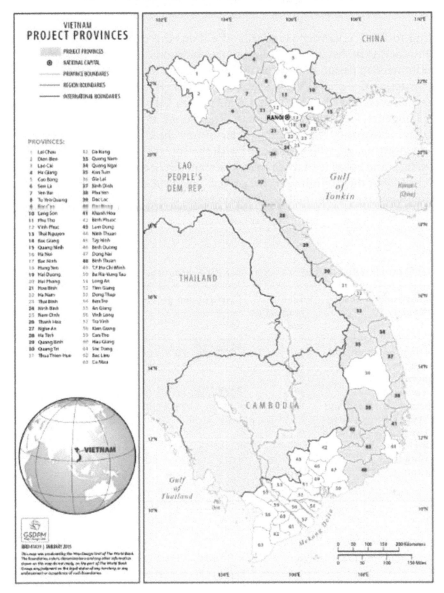

*Figure 8.4* DRSIP – map of Provinces where dams to be rehabilitated are located.

*farmland, but it* also supplies water for industrial and domestic use in Ho Chi Min City. The dam was commissioned in 1985. It has an average height of 28 m, a storage volume of 1,500 MCM and a crest length of 28,100 m.

The main findings from dam safety assessments of the 19 dams are described in the Sections below.

## 8.3.2   Dam classification, design and safety check floods

### 8.3.2.1   Vietnamese standards

According to the Vietnamese standards, the dams are classified in Grades I–IV according to their type, height, and foundation conditions, as shown in Table 8.2 below:

Recommended design and check floods in the Vietnamese standard are based on the Construction Grades I–IV and are given in Table 8.3 below.

It can be seen from Table 8.3, that a maximum Check Flood has a return period of 1 in 5,000 years, for the Special Design Grade dam. It is noted that no downstream consequences have been considered in determining the Check Flood. These floods are largely underestimated, when compared with the International Standards, i.e. ICOLD requirements, and could lead to an unsafe design (see Section 2.4.1). These issues have been picked up at the beginning of the project when the dam classification and floods determination was re-examined (see Section below).

Table 8.2 Dam classification in accordance with Vietnamese standard

| Type of construction and service capacity | Type of foundation soil | Construction grade | | | | |
|---|---|---|---|---|---|---|
| | | Special | I | II | III | IV |
| 1. Earth dam or earth and | A | >100 | >70 ÷ 100 | >25 ÷ 70 | >10 ÷ 25 | ≤10 |
| rockfill dam. with the | B | - | >35 ÷ 75> | >15 ÷ 35 | >8 ÷ 15 | ≤8 |
| maximum height. m | C | - | - | >15 ÷ 25 | >5 ÷ 15 | ≤5 |
| 2. Concrete dam. | A | >100 | >60 ÷ 100 | >25 ÷ 60 | >10 ÷ 25 | ≤10 |
| reinforced concrete and | B | - | >25 ÷ 50 | >10 ÷ 25 | >5 ÷ 10 | ≤5 |
| other types of pressure irrigation structure with the height. in | C | - | - | >10 ÷ 20 | >5 ÷ 10 | ≤5 |
| 3. Retaining wall with the | A | - | >25 ÷ 40 | >15 ÷ 25 | >8 ÷ 15 | ≤8 |
| height. m | B | - | - | >12 ÷ 20 | >5 ÷ 12 | ≤5 |
| | C | - | - | >10 ÷ 15 | >4 ÷ 10 | ≤4 |

Table 8.3 Recommended design and check floods - Vietnamese standards

| Frequency of flood | Construction grade | | | | |
|---|---|---|---|---|---|
| | Special | I | II | III | IV |
| Design flood frequency (%) | 0.10 | 0.50 | 1.00 | 1.50 | 2.00 |
| Year return period (year) | 1,000 | 200 | 100 | 67 | 50 |
| Check flood frequency (%) | 0.02 | 0.10 | 0.20 | 0.50 | 1.0 |
| Year return period (year) | 5,000 | 1,000 | 500 | 200 | 100 |

### 8.3.2.2    Project operation manual

Prior to the commencement of the project, a Project Operation Manual (POM) was prepared to provide the classification of dams and requirements for different dams. The POM also refers to applicable guidelines and standards that need to be applied for dam safety reviews. Requirements from the POM relevant for dam safety review are summarised below.

The POM defines Large Dams by combining requirements of the ICOLD classification and Vietnamese standards, as follows:

*Large Dams are:*

- $H > 15\,m$
- $10\,m < H < 15\,m$, Storage $> 3$ million $m^3$
- $10\,m < H < 15\,m$, Crest length $> 500\,m$
- $10\,m < H < 15\,m$, Spillway capacity $> 2,000\,m^3/s$
- $10\,m < H < 15\,m$, with special design complexities (e.g., large flood handling requirements, location in zone of high seismicity, retention of toxic materials) and a large number of people at risk in downstream area,
- $H < 10\,m$ if they are expected to become large dams during operation

*Small Dams are:*

- $H < 15\,m$. This category includes, for example, farm ponds, local silt retention dams, and low embankment tanks.

*Choice of Design and Check Floods*
The POM recommends the following:

- Use Vietnamese Standards for $V < 3\,Mm^3$ (as per Table 8.3);
- Use POM for Check Floods for Dams with $V > 3\,Mm^3$ (see Table 8.4 below). The number of households at risk is estimated using New Vietnamese "Guidelines on rapid assessment that identify number of affected people at downstream in case of dam failure" (2017).

As it can be seen from Table 8.4, the Check Flood could vary from 1 in 1,000 years to the Probable Maximum Flood (PMF), depending on the downstream population at risk.

*Table 8.4* **Recommended safety check floods - as per POM**

| The number of households in the downstream | Safety Check flood frequency |
|---|---|
| >10,000 | PMF |
| 1,000 ÷ 10,000 | 0.01% ÷ PMF |
| 25 ÷ 1,000 | 0.01% |
| <25 | 0.1% |

### 8.3.3   Reservoir sedimentation

For the Dau Tieng Dam, the largest irrigation dam in Vietnam, with the original storage of 1,500 MCM, it was recommended that the calculated PMF estimates were re-checked as they appeared low; also, the bathymetric survey has never been done for this reservoir which was commissioned in 1985 – it was recommended that this survey was undertaken so that height/storage curves for the reservoir could be updated; all the current flood capacity calculations are based on the original storage figures and largely vary; it is known that sand was mined from the reservoir, therefore sedimentation is clearly an issue.

### 8.3.4   Spillway capacity and safety against overtopping

Many of the 19 dams that were assessed did not have adequate spillway capacity and safety against overtopping. It was recommended that these dams had either:

- the existing spillway widened (Figure 8.5) or modified into a labyrinth weir,
- a completely new emergency spillway constructed (Figure 8.6),
- dam crest raised, and a wave wall was constructed (Figure 8.7).

Some of these works have been implemented in 2018/2019 and checked during subsequent visits.

*Figure 8.5* Widened spillway at Khe Che dam, which was very nearly overtopped during the 1995 flood (the white line shows the location of the side wall of the original spillway).

*Figure 8.6* Newly constructed emergency spillway at Phu Vinh dam. Some dams will have the crest raised and a wave wall added (Figure 8.7).

*Figure 8.7* Dak R'Lon dam – raising of dam crest and construction of wave wall.

### 8.3.5   Seepage through the embankment

For many dams visited seepage through the dam body, foundation, or both, was an issue. Typical anti-seepage measures proposed are the construction of grouting walls. However, for many dams reviewed, the wall did not penetrate sufficiently into impermeable strata; deepening of the wall was recommended on a few dams.

### 8.3.6   Erosion of slopes due to dispersive soil properties

Significant erosion holes were encountered on downstream slopes on several dams (Figure 8.8), which indicates that the homogeneous embankments were constructed of dispersive soils, common in this part of the world. It is recommended to check the dispersive property of the fill by a combination of methods (Crumb, Pinhole, Double hydrometer and chemical) and provide appropriate stabilisation method by lime or gypsum and/or a combination of geotextile filter and stone pitching for the downstream slope;

### 8.3.7   Upstream slope protection

Inadequate upstream slope erosion protection (Figure 8.9) was seen on some of the dams; this can lead to slope erosion and embankment failure and is recommended to be repaired.

*Figure 8.8* Significant erosion holes on downstream slope encountered on several dams.

*Figure 8.9* Inadequate u/s slope protection.

## 8.3.8   Seismic stability of dams

As the current Vietnamese Guideline for the Seismic Design only defines accelerations for 1 in 475 year return period earthquakes, it is recommended to undertake seismic assessments of Large Dams in accordance with requirements of ICOLD Bulletin 148, i.e. check the dams for an OBE and SEE earthquakes; these assessments shall also include checks for susceptibility of the fill and foundation material to liquefaction, as for some dams fill material is largely composed of silty sand.

## 8.3.9   Monitoring of dams

No monitoring instrumentation was envisaged for dams that are classified as Grade III and IV in accordance with the Vietnamese standards – it was recommended that a water level measurement gauge and a "V" notch weir at the lowest point of the drain at the d/s toe are provided as a minimum on these dams.

## 8.3.10   Capacity of local consultants

There is a concern about the capacity of the local Consultants to undertake the PMF estimates, dam break and seismic analysis in accordance with international standards. A possible training should be arranged in these fields.

## 8.4   CONCLUSIONS AND CHALLENGES

Due to the high population density in the country, a demand for water is very high; but also high is a demand to have safe dams as their breach would have a significant impact on many people leaving downstream.

The country is also extremely vulnerable to a climate change, which puts pressure on the capacity of the existing dams to cope with the increased rainfall.

There is a large discrepancy in design floods between the local standards and the international standards; the floods derived from the local standards are evaluated to be underestimated and unsafe, which has already been demonstrated on several dams that have been overtopped in the past or nearly overtopped. This, coupled with the impact of climate change, puts some of the dams in the highest risk category requiring the provision of additional spilling capacity and safety for overtopping.

Attention is needed when using soils prone to dispersion as an embankment fill; adequate investigation and design measures are needed to avoid serious issues during operation.

It is recommended to alter the current standard for dams monitoring and consider the installation of some monitoring instruments on dams classified as Grade III and IV.

Although knowledge of the international standards for dam safety is available to some Vietnamese experts, typically working at universities, training of local consultants is recommended to bring them in line with the safe and sustainable dam design.

## NOTES

1 Flooded futures: global vulnerability to sea level rises worse than previously understood, Climate Centre, 2019.
2 Dam development in Vietnam: The evolution of dam-induced resettlement policy. Dao, N. 2010. Water Alternatives 3(2): 324–340.
3 International small dam safety assurance policy benchmarks to avoid dam failure flood disasters in developing countries, John D. Pisaniello, Tuyet Thi Dam, Joanne L.Tingey-Holyoak, 2015.
4 Dam Safety Project in Vietnam, ICOLD 2020 Delhi.

# Chapter 9

# Case studies

## East Africa – Mauritius

### 9.1 ABOUT THE COUNTRY

Mauritius, shown in Figure 9.1 below, is a volcanic island situated in the Indian Ocean, some 2,000 km off the southeast coast of the African continent, east of Madagascar.

**No. of people:** 1.3 million

**Area** – 2,040 km$^2$

**Population Density** – 618/km$^2$. The population density in Mauritius is the highest of the African countries and is among the highest in the world (tenth most densely populated country in the world). Overpopulation became a serious problem for the country.

**Economy:**

GDP Nominal - (Total) $14.8 billion

GNI Nominal - (per capita) $12,740

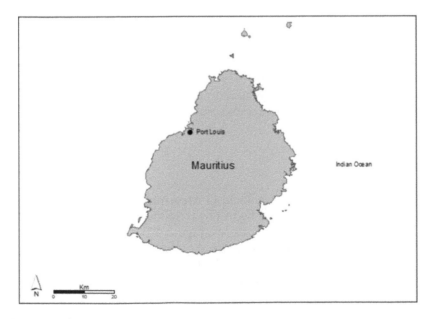

*Figure 9.1* Map of Mauritius.

DOI: 10.1201/9780429320453-9

**Climate** – Mauritius is situated near the Tropic of Capricorn, and the climate stays tropical, with coastal regions but there are mountainous areas with forests. There is a warm and humid summer usually from November to April with an average temperature of 25°C, and it has a cool and dry winter from May to October which an average temperature is around 21°C.

Although the average rainfall over the country is adequate, it has an uneven distribution over the island and irregular seasonal variation (see below). The central area of the island receives up to 4,000 mm of rainfall a year, while the coastal areas receive between 600 mm (Western coast) and 1,400 mm (Eastern Coast) per year. Cyclones can occur between January and March, and heavy rain may occur for an extended period of time.

**Vulnerability to climate change** – Mauritius is amongst the most vulnerable countries to climate change and one of the most exposed to natural hazards due to its geographical location in an active tropical cyclone basin. Mauritius is highly vulnerable to the impacts of intense cyclones, abnormal tidal surges, prolonged droughts, flash floods, and an increase in sea temperature. The country's vulnerability is due to a range of climatic, biological, geological and technological hazards. In recent years, the increasing frequency and intensity of cyclones, torrential rains and flash floods have also threatened people's livelihoods in the islands.

In March 2013, 11 people were killed by flash floods in Port Louis. The passage of Tropical Cyclone Fantala, with 280 km/h gusts, threatened the low-lying islands of Agalega and St Brandon.

In March 2017, the passage of Tropical Cyclone Enawo at 250 km/h gusts in the Southern Indian Ocean region posed a threat to the main island of Mauritius.

More recently, in February 2019, Cyclone Gelena with 165 km/h gusts crossed approximately 50 km southwest of Rodrigues Island leading to flash floods that caused the displacement of 259 people as well as damage to infrastructure, private residences and farms, and severely affected the electricity network. Mauritius' disaster risk profile by the *Global Facility for Disaster Reduction and Recovery* highlights that flooding is the second largest risk after cyclones, causing 20% of the direct economic losses associated with disasters. Most of these costs arise from damage to people's homes.

It has to be noted that rapid urbanisation on formerly agricultural land has strained the national drainage system and increased the occurrence of flash floods, destruction of housing, infrastructure and crops, and putting the population at risk of vector and water-borne diseases.

## 9.2   ABOUT DAMS

Several reservoirs have been constructed in Mauritius since 1885 to improve water supply for domestic, irrigation and hydropower production across the island. There are currently 12 reservoirs in Mauritius (see Figure 9.2) with a total storage capacity of around 95 MCM. The oldest dams on the island are at Mare-aux-Vacoas, La Ferme and La Nicoliere reservoirs, constructed in 1885, 1914 and 1929, respectively, when Mauritius was under British rule. The newest reservoir at Bagatelle, completed in 2018, aims to improve the reliability of supply to the Port Louis and Lower Mare aux Vacoas water supply network.

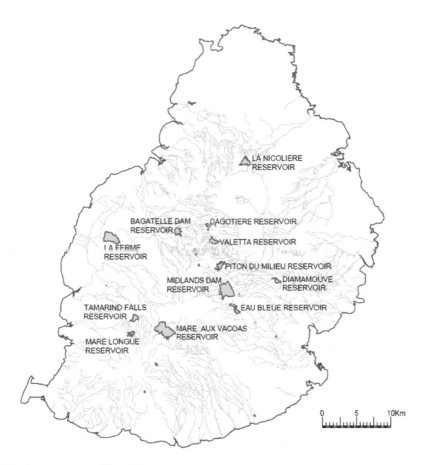

*Figure 9.2* **Reservoirs in Mauritius.**

Due to the increase in population, its density and the demand for water, there is an emphasis on ensuring that the existing dams are safe and that their storage could be augmented to provide an increased supply, without constructing new dams.

**Law and regulations**
No dam specific laws and regulations exist in Mauritius.
**ICOLD membership**
Not a member of ICOLD

## 9.3   LA FERME DAM REHABILITATION PROJECT

### 9.3.1   Project background

La Ferme Reservoir was constructed in 1914 to store water for irrigation purposes. The dam, with a storage capacity of 11.51 MCM, was constructed on the River Belle Isle. The reservoir is fed by surface runoff from its natural catchment (about 25%) and two feeder canals (about 75%), namely the Triannon Grosses Roches Feeder Canal (TGRFC) and La Fenêtre Feeder Canal (LFFC).

Due to decades of operation and ageing, some deficiencies have been identified in the dam, associated structures, and the hydro-mechanical equipment. It has also been reported by the Water Resources Unit (WRU), which operates and maintains the dam and the reservoir, that due to insufficient reservoir's storage capacity, inflows during the wet seasons are often discharged over the masonry spillway and wasted downstream of the dam.

An initial safety assessment of La Ferme Dam was undertaken in 1997 as a part of the safety assessment of seven other dams in Mauritius. A detailed design of safety remediation works was undertaken over 2011/2012.[1] The principal objectives of the La Ferme dam rehabilitation project are:

1. review the safety of the dam and implement rehabilitation measures that would meet the international dam safety standards and extend its useful life;
2. carry out reservoir augmentation studies for increasing the storage capacity, e.g. by raising the existing dam to provide additional surface water resources for irrigation and potential future potable water usage;
3. assess the capacity of the downstream channel to convey the spilt water.

### 9.3.2   Description of the dam

The dam, which extends North-South, is located approximately 15 km southwest of Port Louis, adjacent to the village of Bambous the District of Black River. The reservoir is surrounded by woodland and scrubland with adjacent steeply sided mountains. There are significant numbers of residential dwellings (>100, both legal and illegal) in the immediate vicinity downstream of the dam (Figure 9.3).

Figure 9.4 shows the principal components of the dam, which are as follows:

- The La Ferme Reservoir is 7.4 km in perimeter, with 11.74 MCM usable storage
- Earth Embankments – Right Embankment (RE), 890 m long and Left Embankment (LE) 355 m long. The maximum height of both embankments is 7 m
- Masonry dam (non-overflow and spillway sections), total length 270 m; maximum height 12 m; *the masonry dam comprises an 82 m long right non-overflow section, a 146 m long overflow (broad crested) spillway, and a 48 m long left non-overflow section;*
- Feeder Canals – TGRFC and LFFC
- Bottom outlet to the La Ferme Irrigation Canal and the River Belle Isle

### 9.3.3   Summary of findings based on dam safety assessment

#### 9.3.3.1   Dam hazard classification

La Ferme Reservoir falls into the International Commission on Large Dams (ICOLD)'s INTERMEDIATE dam size and HIGH hazard potential.

*Figure 9.3* **La Ferme dam –a significant number of people live in the immediate vicinity.**

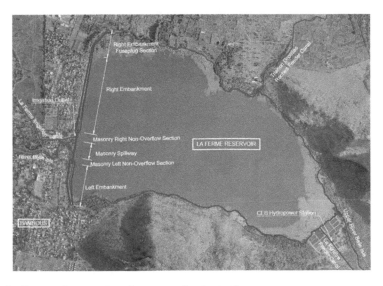

*Figure 9.4* **La Ferme Reservoir – layout and principal components.**

*9.3.3.2   Inspection*

**Right Embankment (RE) and the Left Embankment (LE)**

- the upstream slope is protected with hand-packed 300 mm thick basalt pitching overlaying a 150 mm thick gravel layer (Figure 9.5). The stone protection to the RE, at a location the most exposed to wind and wave action, had been damaged, indicating its inadequate thickness. Additionally, evidence suggests that some damaged areas of upstream face protection are caused by the illegal removal of boulders for building purposes.
- The downstream slope is vegetated so that seepages could be obscured,
- there are many wet areas immediately downstream of the LE, seepage through the foundation in this area has been confirmed during the ground investigation that was undertaken under this project.
- The end section of the RE, 290 m in length, is lower and narrower than the rest of the main embankment which suggests that this area might have been designed to act as a fuse plug, i.e. auxiliary spillway.
- a longitudinal crack occurred several years ago at the RE, which was subsequently backfilled. This longitudinal crack was likely a tension crack, formed over a prolonged dry period due to the high shrinkage/swelling potential of the embankment fill; this has been confirmed during the ground investigation that was undertaken under this project (Figure 9.6).

**Concrete masonry section including the main spillway**

- The masonry dam comprises a 300–400 mm thick stone masonry facing upstream and downstream faces. The body of the dam comprises a cyclopean concrete, which consists of a large range of aggregate sizes from fine to plums (boulder size)

*Figure 9.5* RE stone pitching.

*Figure 9.6* **Springs encountered 80 m downstream of LE.**

contained in a cement matrix to form a roughly homogeneous mix. This reflects a typical masonry construction in the early 1900s.

- Except for some localised damage to some of the facing stones, the dam appeared to be in a good condition, without any signs of movements or leakage.
- However, mature trees are encountered immediately downstream of the masonry dam and spillway (Figure 9.7)

**Irrigation Outlet**

An outlet conduit 0.9 m wide by 0.9 m high on the upstream face, which increases in height to 1.2 m towards the downstream face, has been provided through the right non-overflow section of the masonry dam to supply water to the La Ferme Irrigation Canal (Figure 9.8). The outlet is fitted with a vertical lift penstock at the upstream face. The underwater inspection that was undertaken under this project identified scour damage to the upstream side of the outlet as well as a damaged sealing and generally poor condition of the penstock.

A stilling basin is located at the downstream end of the outlet for energy dissipation, leading to a flow gauging station and into the La Ferme Irrigation Canal. Immediately downstream is a penstock which leads into a channel feeding the River Belle Isle, the assumed purpose of this is to provide compensation flow to the river. The assessment of the existing stilling basin has shown that the basin will not be able to accommodate flows in an emergency situation when the outlet is fully open, and it will need to be modified.

**Feeder Canals: TGRFC and LFFC:**

The TGRFC conveys run-off from catchments to north and east of the natural catchment of La Ferme Reservoir. Flow along the canal is measured close to the

*Figure 9.7* Spillway with mature trees at the downstream toe.

*Figure 9.8* Irrigation outlet.

reservoir via a flow gauging station. The maximum discharge capacity of the feeder canal has been assessed by others to be 2.83 m$^3$/s. It is also proposed that excess spills from the Bagatelle Dam (once completed) will be transferred to the La Ferme Reservoir via the TGRFC.

The LFFC discharges into the reservoir at the east end. The water from the canal passes through a hydropower station before it flows into the reservoir. The maximum discharge capacity of the LFFC has been assessed to be 2.82 m$^3$/s.

### 9.3.3.3  Hydrological assessment

**Yield simulation**

A reservoir yield simulation has been undertaken involving a three-stage approach;

1. Preparation of a mass balance model, using an Excel workbook, to assess the consistency of data sets and produce an 'observed' catchment run-off series.
2. Preparation of a catchment runoff model using the Pitman methodology embedded within WRSM 2000 to produce a synthetic catchment runoff series, calibrated against the 'observed' catchment run-off series.
3. Preparation of a reservoir yield model using the synthetic catchment run-off series and other raw data series.

The managed inflows (LFFC and TGRFC) contribute 75% (approximately 23 Mm$^3$/year) of the total flow into the reservoir, with potential once spills from Bagatelle Dam are made available for increasing to 81% (approximately 34 Mm$^3$/year) of the total flow into the reservoir. The mean annual catchment runoff and direct rainfall onto the reservoir account for approximately 7 Mm$^3$/year – a small proportion of the total inflow. The accuracy of the results of the reservoir yield model is therefore highly dependent on the management policies that determine the allocation of the inter-basin flows.

The mean measured outflow at the irrigation outlet is 14 Mm$^3$/year; however, a measurement error of up to 3.6 Mm$^3$/year is occurring through a scour pipe below the weir at the irrigation outlet.

Excessive seepage is thought to occur at the left embankment and abutment where highly permeable basalt is thought to be exposed with inadequate clay cover. A plausible estimate of total seepage losses is around 3.5 Mm$^3$/year in the existing condition and 5.0 Mm$^3$/year for the raised Full Supply Level (FSL), which is regarded as excessive (see Section 4.1).

The yield/storage analysis does not suggest a significant benefit in raising the crest level of the dam to provide increased storage volume. However, raising the crest of the dam by 1.0 m is estimated to result in an increase in construction cost of only 10% compared with maintaining the current spillway crest. It is therefore recommended that the spillway crest is raised by 1.0–147.72 m. This will provide future operating flexibility to either reduce downstream flood risk or to provide slightly improved yield should future inflows from TGRFC, LFFC and spills from the Bagatelle Dam prove to be reliable.

**Flood Routing, spillway, and River Belle Isle spill capacity**

As the dam is classified to have a HIGH hazard potential, the PMF was adopted for the Safety Check Flood. The peak PMF flow would give a flood rise of 1.70 m. In addition, the reservoir level will be augmented by 1 m to accommodate the future increase in water demand.

The River Belle Isle spill channel immediately downstream of the dam is densely built up through the village of Bambous. The analysis shows that the existing channel has very limited capacity and significant built-up areas are at risk of flooding. Major engineering works to improve the capacity of the spill channel will have limited effectiveness unless realignment and replacement of a major road culvert can also be achieved. It is considered that this level of intervention would pose major social, physical, and financial challenges and is therefore unlikely to be implementable.

It is recommended that some properties immediately downstream of the spillway are relocated. However, relocating all of the at-risk properties is not considered socially, financially or politically acceptable. Therefore, to mitigate the risks to the downstream population it is necessary to develop a dam specific Emergency Preparedness Plan (EPP) together with effective early warning systems and a response matrix.

### 9.3.3.4  Emergency drawdown capacity

It has been assessed that in a case of emergency, the bottom outlet would be able to lower 50% of the reservoir level in 12 days, which is assessed not to be adequate for a dam which has a HIGH potential hazard with a lot of people living in the immediate vicinity downstream.[2]

### 9.3.3.5  Geological and geotechnical assessments

Geological and geotechnical conditions of the site have been assessed based on the published information[3] as well as the results of the Ground Investigation undertaken under this project in 2012.

**Dam Foundations**

Mauritius was formed as a result of undersea volcanic activity and all rocks on the island are volcanic in origin, namely basalts. There have been three distinct series of volcanic activity:

- The Older Series Basalt, Dates 5.5–7.8 Million Years Ago – these rocks have low porosity and low permeability in the order of $10^{-7}$ to $10^{-6}$ m/s.
- The Intermediate Series Basalt, Dates 1.8–3.5 Million Years Ago – permeability varies from $10^{-6}$ to $10^{-3}$ m/s
- The Younger Series Basalt, Dates 0.025–1.6 Million Years Ago – permeability varies from $10^{-5}$ to $10^{-2}$ m/s.

The basalt lavas of Mauritius have been affected by tropical weathering processes. Two different residual soil types that have been identified on the island are:

- Low Humic Latosoil – This soil is formed in the lower rainfall zone, where there is seasonal desiccation of the weathering profile and kaolinite is the dominant clay mineral. The clay is red to brown in colour.

- Humic Latosoil, which forms in the higher rainfall zone, where there is no significant seasonal desiccation. This gives rise to a dark brown/black weathering profile on the basalt lavas. It is black or dark grey clay with Smectite being the dominant clay mineral.

Geology at the dam site comprises young, intermediate and old basalt covered with a layer of superficial deposits, which vary in thickness to up to 3.5 m. The superficial deposits are represented by Low Humic Latosoils and alluvium, in sections closer to the river bed. The clay layer in the dam foundation is very soft in places. At the LE and abutment, the layer of the residual soil is thin and is not present in places. Masonry dam is mainly founded on the Intermediate Series Basalt. A schematic geological model is shown in Figure 9.9.

Based on ground investigation data a seepage model has been established. The seepage analyses performed show that a seepage is likely to occur where the basalt is exposed at the foundation surface; this is in the foundations of the masonry dam and of the end sections of the left embankment and abutment. A plausible estimate of seepage losses is around 3.5 Mm³/year in the existing condition and 5.0 Mm³/year for the raised FSL, which is regarded as excessive. To reduce the seepage through the foundation of the end sections of the LE and left abutment and minimise the water losses, it is recommended to construct a clay blanket, extending some 100 m upstream of the dam.

**Existing Embankments and Masonry Dam**

As no design or as-built drawings were available, the composition of the embankments had to be verified by the ground investigation. Both RE and LE were found to

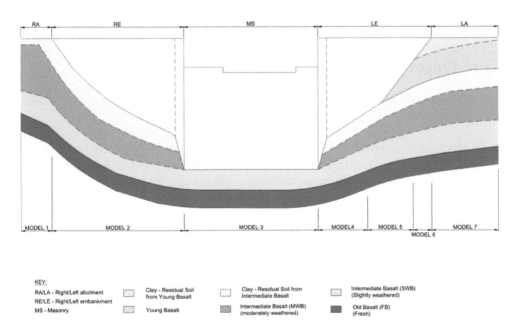

*Figure 9.9* Schematic geological model along dam axis.

be homogeneous, comprising the clay fill. No clay core or a cut-off below the core was encountered during the ground investigation. The LE has layers of granular material within the clay fill and seepage through the embankment has been identified. A phreatic surface within the embankment varies with the reservoir level and is high when the water level is close to FSL.

An old longitudinal crack, subsequently backfilled, was identified along the crest of the RE during site inspection (see (b) above). This indicated that the clay within the embankment fill is likely to be expansive. The ground investigation confirmed that the embankment clay fill has a HIGH degree of expansion/swelling potential. Such clays, if exposed to a prolonged period of dry weather can develop tension cracks. It is likely that the clay from around/within the reservoir was used for the embankment's construction. The investigation confirmed that this clay has HIGH to VERY HIGH expansion/swelling potential.

The investigation confirmed that the concrete core was solid, in a good condition, with no signs of voids or cracks.

### 9.3.3.6   Stability of embankments and masonry dam

The stability of the embankments has been checked and found to be satisfactory with the current slopes.

The stability checks of the masonry dam have also been performed - a particular concern was the impact of the uplift forces on the sliding and overturning, bearing in mind that when the dam was designed in the 1890s, no knowledge of the uplift pressures existed (work by Terzaghi and Casagrande on the uplift pressures originated in 1930s and 1940s). However, the factors of safety on sliding and overturning were found to be adequate.

## 9.3.4   Proposed rehabilitation works

The following rehabilitation works have been recommended to meet the Client's objectives:

- It has been confirmed by the ground investigation that the existing embankment fill and the material in the borrow areas in the vicinity of the dam have high expansive potential. It is therefore recommended that the new fill will either (i) need to be constructed of a clay non-susceptible to swelling/shrinking, or (ii) if the clay from the borrowed area is used, the external clay surfaces will need to be covered with a layer of clay non-susceptible to swelling/shrinking (see Figure 9.10).
- The current stone pitching protection to the upstream slope of the embankments is not adequate and needs to be strengthened by placing a 1 m thick layer of riprap over the existing pitching (see Figure 9.10).
- To reduce the excessive seepage through the foundation of the end sections of the LE and abutment, which comprise the young, highly weathered basalt, it is recommended to construct a clay blanket on the bed of the reservoir, extending some 100 m upstream of the dam (Figure 9.11).

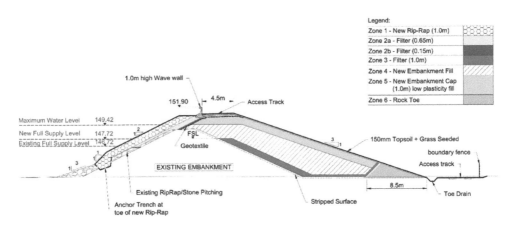

*Figure 9.10* **Proposed typical section through embankment.**

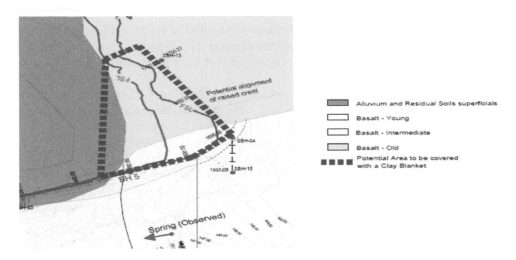

*Figure 9.11* **Proposed upstream clay blanket to reduce seepage losses at the end of LE.**

- To prevent overtopping of the dam in the case of the PMF and wind created waves, the embankment crest will need to be raised by approximately 2.2 m with a provision of a 1 m high wave wall (Figure 9.10). Similarly, the non-overflow masonry section will need to be raised by 2.8 m with a provision of a 1.2 m high parapet wall (see Figure 9.12)
- The existing broad crested spill weir will be modified into a more efficient "ogee" type weir (see Figure 9.12)
- The current bottom outlet does not provide an adequate facility for emergency drawdown. It is recommended that a new, independent, outlet is constructed to facilitate emergency drawdown. This will comprise two 1.0 m ductile iron pipes which will be installed within the masonry section (see Figure 9.12). The pipes will

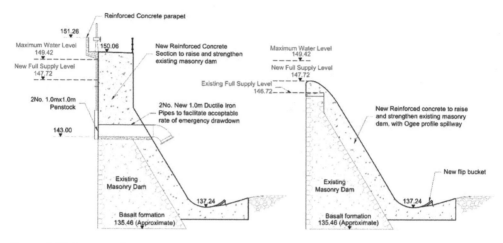

*Figure 9.12* **Proposed section at new emergency outlet (a) and proposed section at spillway (b).**

discharge water over the flip bucket and down into the spillway stilling area. Also, modification to the stilling basin of the existing irrigation outlet will be required.

• The existing sluice gate on the irrigation outlet is in poor condition and will be replaced.
• There are no flow gauging stations to measure the Upper River Belle Isle inflows or the lower River Belle Isle spills and seepage flows. It is recommended that two new flow gauging stations are constructed at these locations.
• Provide dam monitoring instrumentation and O&M manual
• Remove the mature trees at the downstream toe of the dam
• It is recommended that some properties immediately downstream of the spillway are relocated. To mitigate the risks to the downstream population it is recommended to develop a dam specific EPP together with effective EWS and a response matrix.
• Provide a fence along the dam so there is no public access to the dam

## 9.4    CONCLUSIONS AND CHALLENGES

The population density in Mauritius is the highest of the African countries and is among the highest in the world; therefore a demand for water is very high, but also high is a demand to have safe dams as their breach would have a significant impact on a large number of people.

The island is extremely vulnerable to a climate change, which puts pressure on the capacity of the existing dams to cope with the climate change impacts

La Ferme embankments' fill has a HIGH degree of expansion/swelling potential; a deep longitudinal crack was formed in the crest of the embankment which was subsequently backfilled; mitigation measures (protection of crest and downstream slope against wetting/drying) have been recommended). As the dam was designed and constructed over 120 years ago, it is possible that at that time no knowledge of expansive soils and their use as embankment fill existed.

Large water losses through the permeable foundation rock at the end of LE and left abutment were encountered, and mitigation measure (placement of a clay blanket at the upstream toe) has been recommended.

Raising of the dam, provision of a wave wall and modification of the spillway weir is recommended to prevent the overtopping.

Due to the proximity of the people leaving downstream of the dam, there is a need to have the ability to rapidly lower the water in the reservoir in case of emergency – a new emergency outlet structure has been proposed to be provided.

Also, it has been recommended to provide dam monitoring instrumentation, O&M manual, develop an EPP, together with an EWS and the response matrix.

**Challenges**

Design, construction, monitoring and operational records do not exist, which poses a big challenge when assessing the dam's safety performance.

As there are a large number of people living immediately downstream of the dam, it is recommended that some properties immediately downstream of the spillway are relocated. This will be challenging for the government to implement,

Unless there is proper fencing along the dam, with guards, the public might still try and use the dam in an unsafe way.

## NOTES

1 Rehabilitation Works at La Ferme Dam in Mauritius, Spasic-Gril, L., and Wade, D. Water Storage and Hydropower Development for Africa, Addis Ababa, April 2013.
2 Mesures destines a ameliorer la securite des ouvrage hidrauliques des barrages, Combelles et al., 1985; 15th ICOLD, Q59, R46).
3 Soil Erosion Prediction under changing land use in Mauritius, Jacobus Johanne Le Roux, MSc desertion University of Pretoria, February 2005; Directorate of overseas surveys (UK) (1962), Soil Map of Mauritius, Public Works and Survey Department, Port Louis, Mauritius.

# Chapter 10

# Case studies

## East Africa – Madagascar

### 10.1 ABOUT THE COUNTRY

Madagascar, shown in Figure 10.1 below, is in the world's top 50 largest countries, and its neighbouring territories are islands Reunion, Mauritius and Seychelles to the east and Mozambique the nearest mainland African neighbour to the west. There are tropical lowland forests on the eastern side of the island, and to the west, there is a ridge with a plateau in the centre - the altitude ranging from 750 to 1,500 m. The central highlands are the most densely populated area of the island; whereas the west of the island slopes down towards the Mozambique Channel and contains many swamps.

**No. of people:** 28.428 million
**Area** – 587,041 km$^2$
**Population Density** – 48/km$^2$

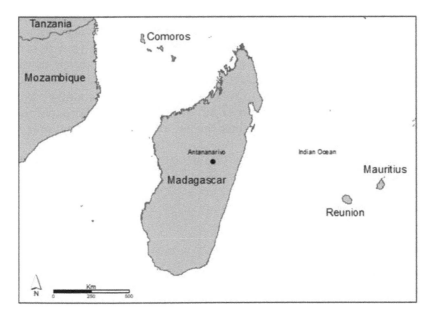

*Figure 10.1* Map of Madagascar.

DOI: 10.1201/9780429320453-10

**Economy:**
GDP Nominal – (Total) $12.73 billion
GNI Nominal – (per capita) $520 – the fourth poorest country in the world
**Climate** – The winds from the southeast and north-western monsoons mean that there is a hot, rainy season with frequent cyclones from November to April and a cooler dry season from May to October. The Indian Ocean produces rain-clouds that add to the moisture on the east coast of the island, and the heavy rainfall contributes to the ecosystem of the country's rainforests. The climate in the central highland is drier and cooler, whereas there is a semi-arid climate in the south and west.

The average annual precipitation varies from 1,000 to 1,500 mm. The coastal region has a tropical climate and no complete dry season. The heaviest rainfall occurs in the coastal region between May and September with average annual precipitation varying from 2,030 to 3,250 mm.

**Vulnerability to climate change** – As Mauritius (see Section 9.1), Madagascar is also vulnerable to climate change and one of the most exposed to natural hazards due to its geographical location in an active tropical cyclone basin.

The tropical cyclones contribute to the damage of infrastructure and economy. In 2004, Cyclone Gafilo was the strongest cyclone on record to hit Madagascar - killing 172 people and injuring further 879.

**Deforestation** – Deforestation by human intervention is significant in Madagascar; it first made an impact on its highland forests as early as AD 600 in the establishment of swidden fields by Indonesian settlers. The creation of *swidden* fields is a subsistence method of agriculture that has been practised by humans across the globe for over 12,000 years by means of a slash-and-burn technique that clears an area in preparation for crop growth. An increase in the rate of forest removal was seen around AD 1,000 with the introduction of cattle from Africa, compelling Malagasy islanders to expand their grassland grazing areas. Historical records point to the importance that this impact has caused with the disappearance of most of Madagascar's highland forest by 1600 AD. Attempts to conserve Madagascar's forests were introduced by rulers in the establishment of environmental regulations, the earliest being seen in 1881 when Queen Ranavalona II placed a ban on using slash-and-burn techniques in agriculture. These efforts aimed to protect the future of the country's rainforests; however, it has been estimated that over 80% of Madagascar's original forests are gone with half of this loss occurring since the late 1950s. In recent decades, trees are cut for exports.

Deforestation has made a very significant impact on soil erosion and sediment transport by rivers (Figure 10.2).

## 10.2   ABOUT DAMS

**Total no. of Dams -14**
   **Law and regulations**
   No dam-specific laws and regulations exist in Madagascar.
   **ICOLD membership**
   Member of ICOLD

*Figure 10.2* Heavily sedimentation of rivers caused by soil erosion due to deforestation.

## 10.3 EMERGENCY PROJECT FOR FOOD SECURITY AND SOCIAL PROTECTION

### 10.3.1 Project background

The National programme »Bassins Versants Périmètres Irrigués (BVPI) is one of the programmes initiated by the Government of Madagascar for the development of the rice sector. Since its start-up phase until 2018, several projects have been set up according to this new approach, including the BVPI/IDA, as well as the Emergency Project for Food Security and Social Protection (PURSAPS), financed by the WB.

These projects all aim to increase agricultural productivity in the large rice basins, where cultivation is done by irrigation.

Several irrigation networks have been built and have been in use for more than 60 years. For various reasons, including insufficient maintenance and the degradation of the surrounding watersheds, rehabilitation works of the irrigation and drainage networks have been planned in these areas.

Under PURSAPS several dams have been assessed for their safety; these are Amboromalandy, Ambilivily and Morafeno dam in the district of Morovoay, region of Boeny, some 500 km north-west of Antananarivo, and Sahamaloto in the region of Alaotra Mangoro, 320 km northeast of Antananarivo. All dams are for irrigation only.

### 10.3.2 Main features of dams

A summary of the dams' main features is given in Table 10.1 below, together with ICOLD (Bulletin 170) size classification.

**Amboromalandy Dam**

The dam was built in 1934 for irrigation. The dam comprises the main embankment and three saddle dams. The main embankment supports the RN4 national road.

*Table 10.1* Characteristics of dams

| Dam | Year of construction | Height (m) | Reservoir volume (MCM) | Size classification (ICOLD) |
|---|---|---|---|---|
| Amboromalandy | 1934 | 8 | 30[a] | Large dam |
| Ambilivily | 1954 | 8 | 13.6[b] | Large dam |
| Morafeno | 1956 | 13 | 13 | Large dam |
| Sahamaloto | 1958 | 14 | 20.6[c] | Large dam |

a   Original volume was 20 MCM; 1 m high parapet wall installed in 1992 and reservoir volume was raised to 30 MCM
b   Original volume was 10.4 MCM; radial gates installed on the spillway in 1992 to raise the reservoir level by 1 m and increase the volume to 13.6 MCM
c   During 1986 rehabilitation the spillway was raised, and the reservoir volume increased by 2.7 MCM to 20.6 MCM

Major dam rehabilitation works were carried out from 1987 to 1992. The works involved:

- construction of the wave wall on the main dam (Figures 10.3 and 10.4)
- placement of an additional berm and a drainage filter on the downstream slope of the main embankment (Figure 10.3),
- construction of a longer, «horseshoe» spillway weir, which replaced the original gated spillway (see Figure 10.5 below)

**Ambilivily Dam**
The dam was built in 1954. The dam is a homogeneous embankment 13 m high. The original spillway, an ogee weir, was modified into a gated spillway during rehabilitation works.

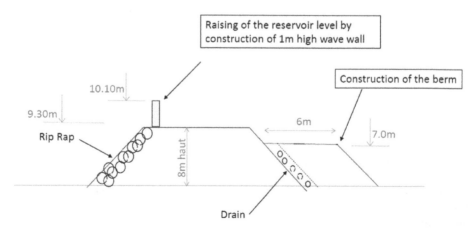

*Figure 10.3* Amboromalandy dam- typical cross section through the main dam, as per 1990 rehabilitation works.

Figure 10.4 Amboromalandy dam- dam crest with a parapet wall constructed in 1990.

Figure 10.5 Amboromalandy dam - new "horseshoe" spillway constructed in 1992.

Rehabilitation works were carried out from 1992 to 1994. The works included:

* raising of the spillway by 1 m by adding radial gates (Figure 10.6); this achieved an increase in volume from 10.4 to 13.6 MCM
* raising the water intake tower

Figure 10.6 Ambilivily dam – raised spillway by addition of radial gates.

Figure 10.7 Morafeno dam- syphon spillway.

### Morafeno Dam

The dam was constructed in 1956. The dam height is 13 m, and the crest length is 700 m. The volume is 13 MCM. There are two spillways:

- the main, syphon spillway (Figure 10.7), and
- The Secondary Spillway – gated spillway with a valve

Intake tower has water intakes at two elevations, each intake is equipped with two gates.

**Sahamaloto Dam**

The dam was constructed in 1958 and comprises five embankments (B1–B4 and b2) with a maximum height of 14 m. The reservoir volume is 20.6 MCM. The spillway is a side channel spillway, adjacent to B4.

During 1983–1986 rehabilitation works, the following were executed:

- raising of embankments crest by 0.6 m to 779.15;
- raising of spillway by 1.6 m to 775.60, thus allowing additional storage of 2.7 Mm$^3$;
- addition of downstream berms and relief wells on B2 and B3 embankments
- construction of a new intake structure at a higher elevation (Figure 10.8)

### 10.3.3    Main findings from dams' inspections

#### 10.3.3.1    Embankments crest

**Amboromalandy Dam**

The dam comprises a main embankment and three saddle dams. Although a parapet wall was added on the main embankment, the crests on the saddle dams remained at their original levels, raising a concern that the saddle dams could be overtopped.

**Ambilivily Dam**

It was noted that although the water level in the reservoir was raised by 1 m by raising the spillway, no embankment raising was undertaken. There is therefore a significant risk of overtopping of the embankment due to a reduced freeboard.

*Figure 10.8* Sahamaloto dam- new intake tower constructed at higher elevation.

**Morafeno Dam**

It was noted during the visit that there was no sufficient freeboard above the FSL, and there is a significant risk of the dam overtopping;

Also, the embankment at its righthand side was previously overtopped at three locations due to insufficient freeboard;

**Sahamaloto Dam**

There is a concern that the embankments are not safe for overtopping.

### 10.3.3.2   Sedimentation

**Amboromalandy Dam**

The reservoir is subjected to sedimentation. The last bathymetric survey was carried out in 1992.

**Ambilivily Dam**

The reservoir encounters sedimentation problems. The last bathymetric survey was done in 1985.

**Morafeno Dam**

No data was available on sedimentation

**Sahamaloto Dam**

Sedimentation is a significant problem for the reservoir. The last bathymetric survey was undertaken in 1998 when the reservoir volume was reduced by 15.2 Mm$^3$. The sediment rate estimated is 840 t/km$^2$/year and with such a rate, it is expected that the current live storage is around 11 Mm$^3$, which would be less than 50% from the original storage.

It was noticed during the site visit that the water was spilling over the spillway, without being retained in the reservoir. There is a concern that the reservoir will not be able to attenuate a major flood and the embankments will be overtopped (Figure 10.9).

*Figure 10.9* Sahamaloto dam – water cannot be retained within the reservoir.

### 10.3.3.3   Seepage at the downstream toe

Seepage at the downstream toe was encountered at Amboromalandy and Morafeno dams (Figure 10.10)

### 10.3.3.4   Spillway gates

**Ambilivily Dam**

It was noticed during the site inspection that cables to some of the radial spillway gates were malfunctioning so not all the gates could be opened (Figure 10.11). This is clearly a dam safety issue as it could lead to dam overtopping.

*Figure 10.10* **Morafeno dam- seepage at the downstream toe.**

*Figure 10.11* Ambilivily dam- cables to some of the spillway gates are malfunctioning and not all the gates can be open.

### 10.3.3.5   Dam access

Ambilivily (Figure 10.12) and Morafeno dam (Figure 10.13) could not be accessed by heavy vehicles after rain, which is not acceptable, in case there is an emergency.

*Figure 10.12* Ambilivily dam- access road to dam crest cannot be used after heavy rain.

*Figure 10.13* Morafeno dam – access road to the dam crest cannot be used after heavy rain.

### 10.3.3.6    Monitoring instruments

There are no monitoring instruments on the dams visited.

## 10.3.4    Dams safety assessemnt

The following has been reviewed as a part of dam safety assessment:

• Hazard classification and choice of safety check floods
• Flood routing, Safety of the Primary and Secondary embankments against over-topping, available freeboard
• sediment transport and deposition in reservoirs vs available storage
• management procedure for the manual operation of the radial gates for the Am-bilivily dam and the emergency radial gate at Morafeno dam.

**Hazard classification and choice of Safety Check Floods**
According to the FEMA 2013 and ICOLD Bulletin 82, the dams have been classi-fied between Low and Significant Hazard Class, with a Safety Check Floods
varying between 100 years to 50% of PMF.
The 100 year, 1,000 year floods and 50% PMF for Sahamaloto only, have been computed and compared with previous studies.
It was found that for Amboromalandy and Sahamaloto dams the recalculated floods are larger than the ones given in the previous studies. For Ambilivily and Morafeno dams the recalculated floods are within boundaries of the floods previously calculated.
**Flood routing Safety of the Primary and Secondary embankments against overtop-ping and available freeboard**
The behaviour of the hydraulic structures to accommodate the check floods was analysed and summarised in Table 10.2. Topo survey was undertaken in 2017 to obtain the exact elevation of the dam crest and spillway.
**Sediment Transport and deposition in the reservoirs**
As the sedimentation was a large concern during the dam safety assessment, the client arranged for a bathymetry survey to be taken as the previous bathymetric survey was done 20–30 years ago.
The results of the 2017 survey were compared with the previous measurements and are presented in Table 10.3 below.

*Table 10.2* Calculation of available freeboard

| Dam | Freeboard (m) |
| --- | --- |
| Amboromalandy | 0.49 (main dam) – insufficient |
|  | • 0.9 (North saddle dam) - overtopping |
|  | • 1.06 (RN4 saddle dam) overtopping |
| Ambilivily | • 0.45 overtopping of main dam (the dam was designed for 100 year flood) |
|  | • 1.15 Overtopping of secondary dam |
| Morafeno | 0.30 (main dam) -insufficient |
|  | • 0.31 overtopping on the right bank of access track |
| Sahamaloto | 0.67 – insufficient |

*Table 10.3* Results of bathymetric survey

| Dam | Live storage volume Measurements | | | | Sediment volume (MCM) | Siltation rate |
|---|---|---|---|---|---|---|
| | Year | Volume (MCM) | Year | Volume (MCM) | | |
| Amboromalandy | 1992 | 30 | 2017 | 24.29 | 5.705 | 19% |
| Ambilivily | 1985 | 13.6 | 2017 | 11.02 | 2.58 | 19% |
| Morafeno | 1992 | 13.0 | 2017 | 11.16 | 1.84 | 14% |
| Sahamaloto | 1998 | 15.2 | 2017 | 8.64 | 6.56 | 42% |

As it can be seen from Table 10.3, sedimentation is a significant problem for all four reservoirs, but the Sahamaloto reservoir is the most affected. The last bathymetric survey was undertaken in 1998 when the reservoir volume was reduced to 15.2 MCM. The sediment rate estimated is 1,240/km$^2$/year, which is very high and leaves just over 50% of the original live storage. The reservoir storage reduction rate: from 1958 to 1992 was 0.135 Mm$^3$/year; from 1992 to 2017: 0.33 Mm$^3$/year, which is a serious increase in sedimentation. It was also noticed during the site visit that the water was spilling over the spillway, without being retained in the reservoir, and there is a concern if the reservoir can attenuate any flood.

Therefore, it has been recommended to assess what would be a suitable remediation option for the dam:

1. Undertake raising of the embankments and increase the capacity of the spillway, or
2. Decommission of the existing dam and construct a new one.

It is envisaged that the raising of the embankments' crest would pose several serious issues that would make that option an unfeasible solution.

## 10.3.5   Summary of findings and further work recommended in the interest of safety

| Dam | Findings in the interest of safety | Recommended work in the interest of safety |
|---|---|---|
| Amboromalandy | Likely overtopping of the saddle dams in places where embankment crest is lower; there is insufficient freeboard at the main dam | • Clearing of slopes, <br> • raising of some parts of the North and RN4 embankments, <br> • provision of a toe drain on the main dam; <br> • Increase flood discharge capacity of the dam (construct a secondary spillway on South saddle dam such that combined capacity of existing spillway and the new one can safely pass the 1,000 year flood); <br> • Provide Instrumentation, O&M, EPP and dam safety equipment |

*(Continued)*

| Dam | Findings in the interest of safety | Recommended work in the interest of safety |
|---|---|---|
| Ambilivily | • Possible overtopping due to insufficient freeboard; construction of a wave wall and raising of embankment in some places; <br> • Urgent replacement of cables to the spillway gates so that they can all be fully operated; <br> • Access road is in an acceptable status that would jeopardise easy access to the crest in the case of emergency | • Clearing of slopes, <br> • construction of a wave wall/ raising of some parts of embankments, <br> • provision of drain at toe, <br> • Urgent works: replacement of cables to spillway gates; <br> • Create a secondary spillway near the existing spillway and increase discharge capacity of the channel of the existing spillway <br> • Repair access road to the dam <br> • Provide Instrumentation, O&M, EPP and Dam Safety equipment |
| Morafeno | • Possible overtopping due to insufficient freeboard, construction of a wave wall and raising of embankment in some places <br> • Access road is in an acceptable status that would jeopardise easy access to the crest in the case of emergency | • Clearing of slopes, <br> • construction of parapet wall/ raising of some parts of embankments, <br> • provision of drain at toe, <br> • Create additional spillway capacity <br> • replacement of cables to auxiliary spillway gates; <br> • Repair access road to the dam <br> • Provide Instrumentation, O&M, EPP |
| Sahamaloto | • The dam has likely reached the end of design life, as the reservoir is sillted up <br> • Embankments are not safe against the overtopping | Study of options (raising or decommissioning of the existing dam & construction of a new dam) |

## 10.4  CONCLUSIONS AND CHALLENGES

Economic capability of the country (the fourth poorest in the world) has been reflected in the dam infrastructure, instrumentation, operation, and maintenance. Extremely limited budgets are allocated for the operation and maintenance of the existing dams.

However, due to the islands exposure to climate change, the country depends on the capacity of dams to store and safely manage the water.

Deforestation, soil erosion, and sedimentation of rivers and reservoirs is a significant problem. In some other catchments, not related to the four dams assessed, the government has already started with the implementation of long term sedimentation management measures, i.e. planting of trees (Figure 10.14) and construction of check dams (Figure 10.15) etc.

The four dams assessed were constructed in the 1930s and 1950s and as such have already passed the design life; all four were rehabilitated in the 1980s and early 1990s, but some rehabilitation works were not adequate (raising of the spillway crest, without

*Figure 10.14* **Androtra region - tree planting in upstream catchment.**

*Figure 10.15* **Androtra region - construction of small check dams in upstream catchment.**

raising of the embankment), so further safety rehabilitation works are required to ensure the dams are not overtopped.

The most serious safety concerns are related to the Sahamaloto dam, which has lost at least 50% of its original storage and would perhaps require decommissioning.

Dams have no monitoring instruments, no O&M and EPP plans which are recommended to be provided.

# Chapter 11

# Lessons learnt and recommendations

Safety assessment of 134 existing dams is presented in Chapters 3–10. The dams are in eight countries, on three different continents: Eurasia, Asia, and Africa, with different political, social, and economic background. The dam safety assessments were carried out from 2002 to date and were a part of various Dam Safety Projects.

The safety of the dams has been evaluated using established international dam safety standards which cover the following aspects:

- Dam safety inspection;
- Robust technical assessment of dam safety to floods, landslides, earthquakes, stability, seepage, etc;
- Assessment of construction details, etc;
- Assessment of risk to the downstream population due to dam breach and dam hazard classification;
- Recommendation of safety mitigation measures that include technical and non-technical measures (monitoring, safe dam operation and Emergency planning and response).

While undertaking these projects, the author realised that many technical aspects were linked to political, social, or economic issues of a particular country; these elements played a major role in various dam safety decisions and have also been highlighted in the book.

Although the dams are located in different countries on three different continents, many of the safety-related findings are common; these are summarised below as they could be lessons learnt that could be applied in other countries as well.

## 11.1 SUMMARY ON THE COUNTRIES

Table 11.1 provides a summary of the information on the countries and portfolios of dams.

### Population density

Myanmar and Vietnam have the largest population of the eight countries, 53,600,000 and 96,000,000 respectively, which explains why the demand for dams is so extremely high. Vietnam has over 7,500 dams and in Myanmar, the number of dams that have been constructed over the last 30 years has tripled.

DOI: 10.1201/9780429320453-11

*Table 11.1 Summary information on the countries*

| Country | Population | Area (km²) | Population density (number of people/km²) | GNI nominal per capita (US$) | Number of dams | Number of dams assessed |
|---------|-----------|-----------|------------------------------------------|------------------------------|----------------|--------------------------|
| Armenia | 3,000,000 | 29,743 | 101 | 4,680 | 92 | 66 |
| Georgia | 3,700,000 | 69,700 | 58 | 4,740 | 50 | 5 |
| Tajikistan | 9,500,000 | 143,100 | 66 | 1,030 | 10 | 2 |
| Sri Lanka | 22,156,000 | 65,610 | 338 | 4,020 | 300 | 32 |
| Myanmar | 53,600,000 | 676,578 | 76 | 1,390 | 200 | 5 |
| Vietnam | 96,000,000 | 331,212 | 295 | 2,540 | 7,500 | 19 |
| Mauritius | 1,300,000 | 2,040 | 618 | 12,740 | 12 | 1 |
| Madagascar | 28,427,328 | 587,041 | 48 | 520 | 14 | 4 |

The population density is the highest in the two island countries, Mauritius and Sri Lanka, with 618 and 338 people/km², respectively. The population density in Mauritius is the highest of the African countries; Mauritius is the tenth most densely populated country in the world. In addition to the increased demand for water, these countries have issues with dams being very close to populated areas as many people are living downstream of the dams.

**Income**

The income in the countries is as follows:

High Income – Mauritius

Upper Middle – Armenia, Georgia

Lower Middle – Sri Lanka, Myanmar, Vietnam

Low Income – Tajikistan, Madagascar; Madagascar is the fourth poorest country in the world

**Dams assessed**

Out of 134 dams assessed, 78 dams have been classified as Large Dams in accordance with the ICOLD's classification.

Most of the dams are between 20 and 80 m in height; however, also included in the evaluation are Nurek Dam and Usoy natural dam in Tajikistan - Nurek dam is 300 m tall and is now the second tallest dam in the world, while Usoy dam is 500–700 m tall, and is the tallest natural dam in the world.

Although some dams discussed in the book are very old (Polonnaruwa dam in Sri Lanka - about 2,500 years old, and La Ferme dam in Mauritius, 120 years old), most of other dams were constructed around 50–60 years ago, with some less than 10 years ago, but with significant problems that can cause serious safety risks.

**Vulnerability to climate change**

Sri Lanka, Myanmar, Vietnam, Mauritius, and Madagascar are all located in the Tropical Zone, between tropics of Cancer (north) and tropics of Capricorn (south). Rainfall patterns in these five countries are similar and are largely influenced by the monsoons, tropical storms and cyclones which regularly cause flooding and problems to the infrastructure.

Tropical cyclones have been more frequent in the last decades causing natural hazards and damages to infrastructure and economy. Vietnam, Mauritius, and

Madagascar are extremely vulnerable to climate change, where several hundreds of people were killed, and thousands were evacuated during the recent cyclones. Infrastructure, including the dams, becomes extremely vulnerable to these extreme weather conditions, which should be taken into account when assessing existing and new infrastructure.

**Deforestation and impact on sedimentation**

Deforestation and soil erosion causing sedimentation of rivers and reservoirs is a significant problem in Madagascar. The issues have been recognised and the government of Madagascar has already started the implementation of long term catchment sedimentation management measures that involve the planting of trees and the construction of check dams.

Sedimentation of rivers is also a problem in Tajikistan and in some areas of Myanmar.

Other countries, where soil erosion and river sedimentation are high, are encouraged to include sustainable catchment management in dam safety programmes; these measures should be a part of the sustainable development of dams' projects.


## 11.2   TECHNICAL ASSESSMENTS

The main aspects and conclusions of the technical dam safety assessments are presented below.

**Design standards**

Although ICOLD promotes dam safety standards and provides guidelines, differences in the design standards applied or dams' operation and maintenance procedures used in these eight countries are still wide.

Armenia, Georgia, and Tajikistan have used SNIPs in the past, and are still mostly using them to date, although there are new Armenian and Georgian Seismic design standards which are mostly in line with the international standards for Seismic design.

Vietnam has its own Dam Design Standards, but some parts are currently under review. It is recommended for Vietnam to review the current standard for dams monitoring and consider the installation of some monitoring instruments on dams classified as Grade III and IV, which, currently, is not required.

Other four countries (Sri Lanka, Myanmar, Mauritius and Madagascar) do not have their own Dam Design Standards but use various international standards, mainly from the UK, France or the US.

**Dam Risk Classifications**

None of the eight countries considers the impact of a potential dam break on downstream population or infrastructure when assessing dam Risk Class for selection of design floods, earthquakes or requirements for maintenance and emergency preparedness.

When evaluating the dams Risk Class in accordance with the ICOLD references, 88 dams out of 134 dams assessed were classified as Extreme Risk Class; it was estimated that, should the failure of these dams occur, over hundreds of thousands of people living downstream of the dams would be at risk.

It is therefore recommended to include downstream impact of a potential dam break when evaluating a dam Risk Class of existing and new dams; this will minimise the risks that the dam breaks could cause to downstream population and infrastructure, as well as rehabilitations that might be needed during operation of dams.

**Design life**

It is the current practice that the service life of a well-designed, well-constructed and well-maintained and monitored dams is expected to be at least 100 years. However, many older dams have been designed to a design life of 50 years, which was a typical assumption for dams designed and constructed in the 1950s and 1960s. A large percentage of the 134 dams evaluated are older than 50 years but are still in operation.

The lifespan of any dam is as long as it is technically safe and operable. In view of the high damage potential of large storage dams, the safety must be assessed based on an integral safety concept, which includes the following elements:

- Structural safety
- Safety monitoring
- Operational safety
- Emergency planning

Therefore, if the proper handling of these safety issues can be guaranteed according to this integral safety concept, a dam can be considered safe. With the number of people living in the downstream area of a dam and the economic development, the risk pattern may change with time, calling for higher safety standards to be applied to the project.

**Design and construction records**

Design, construction, monitoring, and operational records for dams did not exist in several countries; that always posed a big challenge when assessing the dams' safety performance.

**Dam Inspections and Access**

There is inadequate inspection of dams in Mauritius and Madagascar; the inspections are either not done at all, or the records of the inspections are missing. Also, six of 32 dams assessed in Sri Lanka, which are for hydropower only and are owned and operated by the Ceylon Electricity Board, had not been previously inspected and no inspection records were available. All six dams are Large Dam, with the Extreme Risk Class. This situation is not acceptable and the clients in these countries have been encouraged to implement procedure for dam inspections and monitoring.

Accesses to some dams in remote areas of Armenia, Tajikistan, Myanmar, and Madagascar pose a serious challenge. In Myanmar and Madagascar, the access to some of the dams after floods is not possible; this needs to be improved as it could cause large problems if an emergency coincides with floods. Some on - site-based emergency equipment has been recommended at these, hardly accessible dam locations.

**Design Floods**

For all dams evaluated there is a large discrepancy between the floods adopted in the original designs and the ones recommended by international standards; the floods adopted in the designs were underestimated and unsafe. This, unfortunately, has already been confirmed on several dams that have been overtopped or nearly overtopped in the past. The most critical are design floods used in Vietnam, based on the local standards.

The underestimation of the design floods, coupled with the impact of the climate change, has put some of the dams in the highest risk class, requiring the provision of additional spilling capacity and other safety measures to prevent overtopping (see below on spillways and freeboards).

It is therefore recommended to check the domestic versus the current international standards when selecting floods for dam safety assessment.

**Sedimentation**

Many rivers in Madagascar and some rivers in Sri Lanka, Myanmar and Vietnam have high sedimentation rate.

The high sedimentation rates were the main concern for dams in Madagascar; with the Sahamaloto reservoir being the most affected. The water offtakes in the original water intake tower became blocked by the sediments and the tower had to be replaced by a new one, which was constructed at a higher elevation. There are currently issues with the operation of the gates in the new intake tower as well. It was also noticed during the site visit that the water was spilling over the spillway, without being retained in the reservoir. As there is no attenuation of the floods within the reservoir, there is a concern about the discharge capacity of the existing spillway; one of the options for this dam is for it to be decommissioned.

Dam C in Myanmar, constructed 5 years before the safety assessment was undertaken, showed heavy sedimentation; the sedimentation rates observed were three times higher than the originally estimated. Based on the bathymetric survey undertaken 5 years after construction, it was found that. All the dead storage has been filled with sediments in less than 5 years of operation; the sediments could soon impact the operation of inlet gates for the main canal; for this reason, two additional canals have been constructed, with high-level off-takes.

It is therefore essential that the sedimentation rates are correctly estimated, especially for the existing dams which have been designed for a design life of 50 years but have been in operation longer than 50 years.

It has also been recommended to undertake bathymetric surveys to assess the actual sediment yield and check the available capacity of the dead/live storages for dams in Madagascar and some of the dams in Sri Lanka, Myanmar, and Vietnam.

Also, implementation of sustainable catchment sediment management measures, where appropriate, is recommended to prolong the lifespan of the reservoirs.

**Geotechnical conditions**

Information on geotechnical conditions of the site and materials used for dams' construction was not available for many dams examined.

For some more recently constructed dams, like the ones in Myanmar and Vietnam, the information on the ground investigation was available but it was found to be inadequate. Very often the investigations undertaken were insufficient due to limited budgets and programme constraints; this situation should be avoided as it could lead to inadequate designs and construction.

For most of the dams in Myanmar and Vietnam, the embankments were constructed of dispersive soils, which are common soils in that part of the world. However, soils' dispersive potential was not adequately investigated prior to design and construction, and the designs produced were unsafe as they did not provide specific measures when dealing with dispersive soils.

In Mauritius, the embankments of La Ferme dam were constructed from expansive clays, which were not adequately protected to prevent their shrinking and cracking. That caused a safety problem and needed to be rectified.

Hence, due to inadequate ground investigations, a lot of damages to the embankment slopes have been identified on dams in Myanmar, Vietnam and Mauritius which require comprehensive rehabilitation works.

### Seismic Design Parameters

Several countries, such as Armenia, Georgia, Tajikistan and the northern part of Vietnam are regions that are prone to strong earthquakes.

Dams assessed in these countries, as well as in Myanmar, have been designed to local Seismic Standards which provide seismic coefficients for earthquakes of a return period of 475 years.

Several ICOLD bulletins offer the best practice for seismic evaluations, selection of seismic parameters and seismic observations of dams. ICOLD recommends assessing dams for two earthquakes, namely SEE and OBE. Where appropriate, the Reservoir Triggered Earthquakes shall be estimated as well.

To define these seismic design parameters a site-specific probabilistic seismic hazard assessment is recommended to be carried out to best international practice.

### Spillways and freeboard

The main safety issue on most of the dams was the discharge capacity of the spillways; the spillways were typically under designed and needed major upgrades, where the higher floods could not be sufficiently attenuated in the reservoir. For several dams, where it was possible, construction of additional, emergency spillways was recommended.

There are several dams for which raising of the crest and/or installation of a wave wall were recommended to ensure sufficient freeboard is provided against overtopping.

It is therefore important to assess the design and safety check floods by using the most up-to-date international standards, which take into account the dam risk classification and impact of climate change, as discussed above.

### Emergency Drawdown Facilities

Capacity of spillways and outlet works to adequately lower the reservoir down in a case of an emergency has been checked. When the spillway and bottom outlet were inadequate and could not lower the water level at the required drawdown rate in a case of an emergency, it was recommended to construct a new emergency outlet (La Ferme dam in Mauritius). This is especially critical when many people live immediately downstream of the dam.

### Instrumentation and Monitoring Plans, Operation and Maintenance Plans and EPPs

Only on a small number of dams in a few countries monitoring instrumentation exists. This is also true for the O&M Plans and EPPs.

Typically, a negligible budget for monitoring and maintenance of dams was provided. However, the perceived shortcomings in present O&M procedures are as much the product of inadequate budgets and the failure of management to recruit, train and financially reward staff of the calibre necessary to operate and maintain dams, as they are deficiencies in management procedures and practices. This may be seen as being a failure by Governments to recognise the importance of the security of the nation's stock of dams to the national economy, and the threat that unsafe dams pose to the public at large. These aspects need to be addressed by the governments.

### Portfolio Risk Assessment (PRA)

With an increasing number of dams ageing, and population growth, dam safety has been seen to be an important factor for safeguarding human lives and properties of the people living downstream of the dams, as well as the infrastructure. This applies to the existing dams in the eight countries studied, but also to many other countries.

The PRA has been proven to be a powerful tool which dam owners can use to manage downstream risks and prioritise the rehabilitation works for dams. The PRA methods that can be adopted depending on the amount of data the dam owners have at the time of the analysis.

In the projects presented, the use of the semi-quantitative risk portfolio tool for undertaking the PRA played an essential role in the prioritisation of the implementation of remedial measures.

The PRA was also used to estimate the cost-effectiveness and impact of structural and non-structural measures.

**Dam safety rehabilitation works**
Dam safety rehabilitation works were grouped into:

* structural measures,
* non-structural measures

The structural measures typically involved increasing spillway discharge capacity, provision of adequate freeboard, enhancement of slopes stability, especially in seismic conditions, implementation of measures against dispersive and expansive soils, provision of drainage works and upstream blanket to reduce seepage losses etc.

Non-structural measures included the installation of monitoring instruments and preparation of monitoring plans, flood forecasting, preparation of O&M plans, provision of the O&M equipment, improvements of access roads for safety interventions, preparation of EPPs and procurement and installation of telecommunication equipment, together with sirens.

Dam Safety Projects undertaken in all eight countries demonstrated that by the implementation of both structural and non-structural measures, a significant safety improvement would be achieved for the dams. It was also demonstrated that implementation of effective EPPs alone, for the Extreme Risk Class dams, would play a significant contribution towards the safety of the population living downstream.

## 11.3  POLITICAL AND SOCIAL ELEMENTS

**Armenia, Georgia and Tajikistan**
Until 1991 Armenia, Georgia and Tajikistan were a part of the Soviet Union. They all became independent countries in 1991. These political circumstances had some implications for dams safety in these countries.

*Working with SNIPs*
Local engineers in these three countries have traditionally worked within the Soviet design standards, SNIPs, rather than Western norms, which often impacted the practicality and progress of the design of remedial works, site investigations or laboratory testing and caused delays if other investigations or testing were specified as a part of dam safety assessments. This was also a challenge for the projects that were prepared for the implementation of remedial works, which all had to get approvals from various local committees that only worked within the SNIP norms.

*Reliable information on dam design and initial operation*

It proved impossible to obtain reliable information on dams in Armenia and Tajikistan, due to the aftermath of the collapse of the former Soviet Union. A lot of original investigations and design data, originally owned by the Soviet Institutes, were not made available to the local clients.

*Operation and maintenance of dams post the independence*

All three countries suffered economically post the independence; not sufficient funds were available for the dams' maintenance and operation, so many monitoring instruments have been broken and not been repaired or replaced.

In Armenia, the break up of the Soviet Union and the civil war that followed, put the unfinished Marmarik and Kaps dam at risk, due to a lack of sufficient funds to complete the dams - Marmarik dam was only rehabilitated in 2012; construction works on the Kaps dam rehabilitation are planned to start in 2022.

The political situation in Tajikistan after independence played a significant role in dams' operation; Nurek dam was designed, constructed and initially operated by the Soviet engineers; Lake Sarez and Usoy dam were investigated and monitored by the Soviet engineers. On both sites, after the break up of the Soviet Union, both dams were left in hands of local engineers, who were, initially, not sufficiently trained.

*Transboundary issues*

The implication of potential failures of both Nurek dam and Usoy natural dam in Tajikistan would be catastrophic; in a case of a dam break, both Nurek reservoir and Lake Sarez would drain into the Pyanj River, which is a transboundary river.

Therefore, the safety of both dams has been a priority for the Tajik Government for the last two decades. Although this might have not been such an issue when Tajikistan and other downstream countries were a part of the Soviet Union, with the independence, a potential dam break and flooding in other countries would have large international implications. Therefore, most of the dam safety works and assessments have been underlined by the issues of appropriate monitoring, flood forecasting, provision of instrumentation, and emergency preparedness. Several trainings were provided to the clients in Tajikistan on the transboundary issues and mitigation measures and procedures.

**Political pressure for rapid dams' development**

A rapid increase in the design and construction of dams in Myanmar in the last three decades has put a pressure on dam designers and contractors to deliver projects quickly and economically; all the dams assessed had underestimated floods and sediment yield, and they all had the same typical cross sections with a lack of appropriate anti-seepage measures and absent monitoring instruments. Also, the ground investigation that was undertaken prior to or during construction was not adequate to detect geotechnical issues.

These gaps in design and construction led to serious dam safety risks and a need for safety mitigation measures less than a decade from the dams' construction. The clients and governments should recognise these issues and adopt adequate programme and budgets for safe design, construction and monitoring of dams.

**Dam safety legislation**

When the safety assessment of dams was undertaken in eight countries there were no specific laws and regulations related to dams' safety in the countries.

Since the Dam Law was enacted in Myanmar in 2015.

In Sri Lanka, after many internal discussions under Dam Safety Project, the Client decided that Sri Lanka is not ready for legislation and that the proposed provisions should be contained in a Code of Practice. This is still in place to date.

There is clearly a need for a regulatory function and promotion of Dam Safety Units within the national government ministries for better oversight of dam safety management.

As only Myanmar has the Dam Law enacted, the Dam Safety Units could have a regulatory function, including the task of setting standards, making periodic inspections, conducting the evaluation of the safety status, issuing notices of deficiencies to the owners, followed by monitoring of the corrective work performed. The regulator should also review the designs and construction of new dams to assure that the necessary safety features are included.

**Training and institutional strengthening**
The dam safety performed in several countries highlighted issues in:

* sustainable design knowledge
* appropriate maintenance and operation
* emergency preparedness

These issues were discussed with the clients, who were open to staff training and capacity strengthening, which have since been undertaken in several countries.

It is important to highlight that there is a clear need for capacitating different dam safety target groups at the national level including Dam Owners, Regulatory Authorities, Dam Operators, Policy Makers, Member of Parliaments, Emergency Response Units, Consultants, High Learning Institutions, Private Sectors responsible for dam safety management, NGOs, etc. These target groups are responsible for dam safety from designs and construction of new dams to assure that the necessary safety features are included.

It is recommended to take the above into account when scoping dam safety training and capacity building for different projects.

# Index

154   Index